T0275721

Energy Storage for
Sustainable Microgrid

Energy Storage for Sustainable Microgrid

David Wenzhong Gao
University of Denver, USA

AMSTERDAM • BOSTON • HEIDELBERG • LONDON
NEW YORK • OXFORD • PARIS • SAN DIEGO
SAN FRANCISCO • SINGAPORE • SYDNEY • TOKYO

Academic Press is an imprint of Elsevier

Academic Press is an imprint of Elsevier
125, London Wall, EC2Y 5AS
525 B Street, Suite 1800, San Diego, CA 92101-4495, USA
225 Wyman Street, Waltham, MA 02451, USA
The Boulevard, Langford Lane, Kidlington, Oxford OX5 1GB, UK

ISBN: 978-0-12-803374-6

British Library Cataloguing-in-Publication Data
A catalogue record for this book is available from the British Library

Library of Congress Cataloging-in-Publication Data
A catalog record for this book is available from the Library of Congress

For Information on all Academic Press publications
visit our website at http://store.elsevier.com/

Working together
to grow libraries in
developing countries

www.elsevier.com • www.bookaid.org

CONTENTS

For additional information on the topics covered in the chapter 4, visit the companion site: http://booksite.elsevier.com/9780128033746/

FOREWORD

The modern power grid is one of the most complex man-made engineering systems delivering close to 1000 GW of electricity in the United States alone. Power generation in the traditional power grid is highly centralized with power and energy flowing unidirectionally from synchronous generators through a transmission/distribution network to end users. However, technological issues of traditional electric utilities as well as environmental problems caused by the combustion of fossil fuels have stimulated the research and development of new power system technologies. With the emergence of distributed energy resources (DER), for example, wind, photovoltaic, battery, biomass, microturbine, fuel cell, etc., microgrid technologies have attracted increasing attention as an effective means of integrating renewable distributed generation (DG) into power systems.

A high level penetration of renewable energy resources (e.g., wind, PV) in microgrids can make maintaining grid stability and delivering reliable power challenging due to intermittency and fluctuation issues. In such cases, a distributed energy storage (DES) can play an essential role in improving stability, strengthening reliability, and ensuring security. This monograph is dedicated to fundamentals and applications of energy storage in renewable microgrids. With limited page budget, this book covers the following topics, which are summarized in the following paragraphs: basic concepts and control architectures of microgrids; applications of energy storage systems (ESS) in renewable energy microgrids; interfacing between ESS and microgrid; coordinated frequency regulation of battery energy storage systems (BESS) with renewable generation in microgrid; and sizing of ESS for microgrids.

Nowadays, DG technology is becoming increasingly mature, and is deployed as active distribution networks working cooperatively with conventional power grids. In addition, the issues of exhaustible natural resources, fluctuating fossil fuel prices, and the security of electricity have encouraged governments around the world to hold positive attitudes toward the development of emerging microgrids. Future microgrids will allow high renewable penetration and become building blocks of smart grids thanks to advanced communication and information technology. As the underlying scientific and engineering research questions are being answered, there is no doubt that microgrids will play an extremely important role in future sustainable power and energy systems.

There are several applications of ESS including aggregated and distributed ESS in renewable energy microgrids. The microgrid energy management system includes load leveling and peak shifting features, which are widely used to mitigate load fluctuations and improve power quality. ESS is typically used to suppress fluctuations in renewable sources, with methods such as constant power control, output filtering and ramp-rate control. Uninterruptible power systems (UPS) are another important application of ESS in microgrids, especially for the islanded renewable microgrid. ESS in a microgrid also provides benefits for power quality, voltage regulation, reactive power support, and operating reserves.

Interfacing circuits are needed for an ESS to connect to the microgrid. It is beneficial to provide an overview of structures and basic principles of several power converters such as DC-DC converters, AC-DC rectifiers, DC-AC inverters, AC-AC converters. This is done in Chapter 3. The most important DC-DC converter for an ESS is the bidirectional buck-boost DC-DC converter, which is responsible for the charging and discharging of ESS. For DC-AC converters, the voltage source inverter (VSI) is the most widely used converter in practice. A VSI can be used to integrate an ESS or solar photovoltaic into the microgrid. With dq control method, real power and reactive power are controlled independently. There are different configurations of battery management systems (BMS). Within a BMS, cell balancing is important for reliable operation of the BESS.

Compared with frequency regulation by wind generation system, a BESS is a better alternative for providing frequency regulation and inertial response in a faster, more accurate and flexible manner. So, participation of BESS in an islanded microgrid frequency regulation can assist renewable DGs in operating at their maximum efficiency without excessive power curtailment. In Chapter 4, coordinated frequency regulation of BESS with renewable generation in an islanded microgrid or microgrid clusters is discussed. The objective of microgrid frequency regulation is to regulate the frequency of an islanded microgrid to the specified nominal value in the event of frequency disturbance, and at the same time to maintain the tie-line power interchange among different microgrids within a microgrid cluster, or between two virtual areas within a single islanded microgrid at the scheduled value by coordinating the outputs of wind power generation and BESS through virtual inertial response, frequency droop control, and load frequency control.

Energy storage sizing is an important aspect of the cost-effective functioning of microgrids. In the last chapter, different ESS sizing technologies are evaluated. Cost-benefit analysis is a very common method to determine optimal storage sizing. The implementation of an expansion planning method for optimal storage sizing is included. The objective of this method is to minimize the operating cost, maintenance cost and investment cost of the entire microgrid system. In the case study, an optimization problem for determining both the optimal power rating and energy rating of ESS in a microgrid is formulated and solved with mixed integer linear programming (MILP).

I would like to thank those who have provided help and support during the preparation of the monograph. I am grateful to members of the Renewable Energy and Power Electronics Laboratory at the University of Denver, who have devoted a lot of effort and assistance during the book preparation. Special thanks go to these members: Ibrahim Alsaidan, Xiao Kou, Ibrahim Krad, Qiao Li, Siyang Liao, Shruti Singh, Ziping Wu, Weihang Yan. My thanks also go to all the reviewers and staff members of Elsevier for their timely efforts and support. Last but not least, I would like to appreciate the constant support and guidance from Professor Bikash Pal of Imperial College London.

CHAPTER 1

Basic Concepts and Control Architecture of Microgrids

1.1 INTRODUCTION

This chapter discusses the basic concepts and control structures of microgrids. Nowadays, distributed generation technology is becoming more and more mature, and is deployed as key elements of active distribution network working cooperatively with conventional power grids. In addition, the issues of exhaustible natural resources, fluctuating fossil fuel prices and security of electricity have encouraged governments around the world to hold positive attitudes toward the development of emerging microgrids. Future microgrids will allow high renewable penetration and become building blocks of smart grids thanks to advanced communication and information technology. As the underlying scientific and engineering research questions are being answered, there is no doubt that microgrids will play an extremely important role in future electric power and energy systems.

1.1.1 Concepts of Microgrids

Power generation in the traditional power grid is highly centralized, with power and energy flowing unidirectionally from large synchronous generators through a transmission/distribution network to end-users. However, the technological issues associated with traditional electric utilities, as well as the environmental problems caused by the combustion of fossil fuels, have stimulated research and development into new power system technologies. With the emergence of distributed energy resource (DER) units, e.g., wind, photovoltaic (PV), battery, biomass, micro-turbine, fuel cell, etc., microgrid technologies have attracted increasing attention as an effective means of integrating such DER units into power systems. However, there is no clear definition of a microgrid, and the concept varies in different countries and regions. Based on the European Technology Platform of Smart Grids [1], a microgrid is a platform that facilitates the integration of distributed

Energy Storage for Sustainable Microgrid. DOI: http://dx.doi.org/10.1016/B978-0-12-803374-6.00001-9

generators (DG), energy storage systems (ESS) and loads to ensure that the power grid can supply sustainable, price-competitive and reliable electricity. Figure 1.1 shows a typical microgrid structure, comprising DGs, such as combined heat and power unit (CHP), micro-turbines, PV systems, wind power systems, fuel cells; a distributed energy storage (DES) facility such as battery banks, super-capacitors, flywheels, electric vehicles; flexible loads and control devices.

Microgrids can be classified as AC and DC types. AC microgrids can be integrated into existing AC power grid, but they require quite complicated control strategies for the synchronization process in order to preserve the stability of the system. On the other hand, DC microgrids have better short circuit protection and significantly improved efficiency. Furthermore, some synchronous units (e.g., diesel generators) and some non-synchronous units (e.g., micro-turbine machines) are usually connected in the same microgrid system. As the penetration level of more DC loads (especially Plug-in Hybrid Electric Vehicles) increases, hybrid AC/DC synchronous/non-synchronous microgrids via multiple bi-directional converters will become increasingly attractive. Figure 1.2 shows a typical system structure for a hybrid AC/DC microgrid that contains power electronic interfaces and multiple DER units.

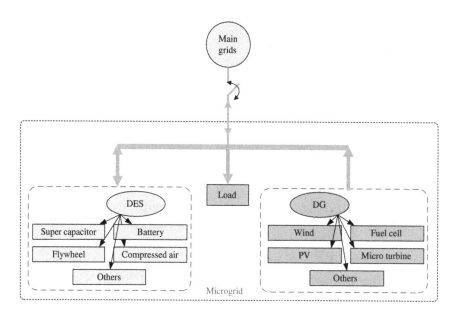

Figure 1.1 Typical structure of a microgrid.

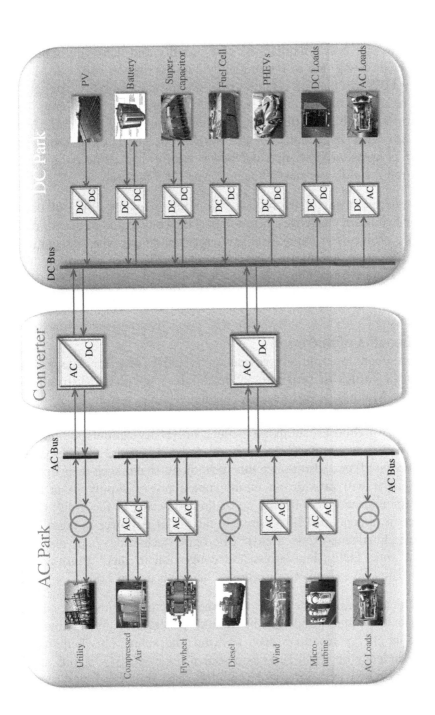

Figure 1.2 Typical system structure for a hybrid microgrid.

Although many types of DG units are more sustainable, a high level penetration of renewable energy resources (e.g., wind, PV) in microgrids can make maintaining grid stability and delivering reliable power challenging due to intermittency and fluctuation issues. In such cases, a DES can play an essential role in improving stability, strengthening reliability, and ensuring security. Not only can DES units be used for smoothing the fluctuations from the output of DG units, but they can also contribute to the stable operation of microgrids. Advances in material science and power electronics technologies have facilitated the effective employment of new DES facilities.

The development of microgrids will bring many benefits but does present significant challenges. For instance, the voltage and frequency disturbance problems in unpredictable weather conditions when integrating renewable energy, monitoring and managing local power generation and loads, designing protection devices to cope with bi-directional power flow and so on. More research needs to be conducted to solve these problems.

1.1.2 Benefits of Microgrids

As mentioned, microgrids provide an effective way for integrating small-scale DERs in proximity of load into low-voltage distribution network. Microgrids can supply highly reliable power to a wide range of customers, both residential and commercial, such as schools, hospitals, warehouses, shopping centers, university campuses, military installations, data centers, etc. Various research stations (Arctic-based or space-based) can also utilize this technology to enhance their operation since it will provide an uninterrupted power supply. It is also useful for remote places having no or limited access to the utility grid. Further, it is beneficial for customers facing large power outages (for example, hurricane-prone areas). Microgrid technology can also be used in areas facing high stress and congestion in their transmission and distribution systems (for example, the northeastern US).

There are many benefits of implementing microgrids. They help facilitate the integration of distributed generation, most notably, renewable energy resources such as wind and solar. This helps curb the dependency on fossil fuels as a source for electricity, significantly reducing carbon emissions and pollution, and thus promotes energy sustainability. They also facilitate the use of highly efficient generators which utilize combined heat and power technology. They can increase the quality of power at the consumer side. With proper control, micro-grids increase electrical reliability by decreasing outage occurrences as well as their duration. Utilities see microgrids as controllable loads, which can contribute to peak shaving during times of peak demand by reducing their own consumption via shedding of non-critical loads and delivering more power to the main grid utility. Microgrids can lower overall distribution system losses by implementing distributed genera-tion located at the demand site eliminating the need for transmission lines and deferring the construction of new transmission lines to a later time. This also results in higher energy efficiency. By using renewable energy resources like wind and solar fuel costs can be reduced. There are also several economic opportunities for microgrids if they can participate in local electricity markets. They can offer several ancillary services to the main grid if properly incentivized to do so. Microgrids can provide active power support via frequency regulation, black start support, system restoration support, and load balancing services. Microgrids can be compensated for these services via fixed payments, payments for service availability, payments based on frequency of usage, and/or payments based on lost opportunity cost. This last is the revenue that the microgrid could have made but was not able to because it had to be available for the main utility grid even if it was not called upon [2].

1.1.3 Integration of Microgrid to Distribution Networks

Conventional DGs are usually directly interconnected to distribution networks at medium or high voltage levels. However, generators in microgrids (e.g., PV, wind turbines, fuel cells) have a relatively small installed capacity (e.g., a few hundred kWs). These generators should be connected to distribution networks at a low voltage level. In conventional power systems, loads are passive and power only flows from distribution substations to customers, but not in the opposite way. But power can flow in both directions between microgrids and the main grid.

In the US, the Federal Energy Regulatory Commission (FERC) provides oversight for constructing electric generation, transmission or distribution facilities. FERC permits various ways of integrating renewable energy resources to facilitate electricity market reform.

The technical requirements for distribution interconnections have been stipulated in IEEE 1547 "IEEE Standard for Interconnecting

Distributed Resources with Electric Power Systems". IEEE 1547 is suitable for all distributed resource technologies, with aggregate capacity of 10 MVA or less at the Point of Common Coupling (PCC).

1.1.4 Basic Components and Operation Strategies in Microgrids

The controllable components in a microgrid include renewable sources, dispatchable sources, ESS and demand side management. All of them work together to maximize the total microgrid profit.

The load control scheme in microgrid can either run in non-automated or automated mode. In non-automated mode, customers can obtain the electricity price and choose whether to switch on or off the loads via remote controls. In automated mode, on-off control is realized by loads themselves through pre-programming or receiving control signals.

The objectives of operating a microgrid depends on the stakeholders' interest. These stakeholders could be microgrid operators, distributed generation owners, distributed generation operators, consumers, etc. To maximize the economic profit, the objective is to minimize total microgrid costs, taking into consideration the impact of the microgrid on the main power grid and the environment. To maximize the technical profit, the objective is to minimize the total power losses and voltage fluctuations, and this option has been adopted by majority of system operators. To maximize the environmental benefits, the objective is to minimize the emissions from the DG in order to meet environmental requirements. The final goal is to combine all the above economic, technical and environment factors to achieve maximum comprehensive benefits [3].

1.1.5 Microgrid Market Models

The microgrid market model consists of consumers, distributed generation owners, the market regulator, retail suppliers, energy service companies (ESCO), distribution system operators (DSO) and microgrid operators. The motivation for using microgrids can be analyzed from either the distributed generation side or the demand side.

On the distributed generation side, since most governments around the world encourage the development of sustainable and clean energy, there are no strict rules for controlling the amount of power output of renewable energy system units. The distribution system operator is responsible for accepting all the electricity from the microgrid if its integration does not impact the safe operation of the grid [3].

On the demand side, the demand response can be classified by the way that the load changes. The first demand response is price-based and the second is incentive-based. In the first situation, the demand response is based on the real-time pricing, critical-peak pricing and ToU (Time of Use) rates. In the second situation, the program operator can switch on or off the customers' loads remotely without considering whether the electricity price is fixed or changing. The purpose of employing these two demand response strategies is to reduce the difference between the load peak and the load valley.

In addition to this, microgrid also provides ancillary services based on the specific requirement of transmission or DSO. The goal is to maintain system stability as well as improving its power quality. Ancillary services can also be divided into two categories according to the operation modes of microgrids:

- Grid-connected mode:
  Frequency control service, voltage control service, grid losses reduction service and power quality service.
- Islanded mode:
  Black start service, frequency control service and voltage control service.

1.2 MICROGRID CONTROL ISSUES

The most important feature that distinguishes a microgrid from a conventional distribution system is its controllability, the purpose of which is to make microgrids behave as a controllable, coordinated module when connected to the upstream network.

1.2.1 Introduction

The function of microgrid control can be divided into three parts; the upstream network interface, microgrid control and protection, local control [3].

The upstream network interface decides whether the microgrid is able to operate in grid-connected mode or islanded mode. It makes decisions for market participation and coordination with the upstream network. The microgrid control includes voltage and frequency regulation, real and reactive power control, load forecasting and scheduling, microgrid monitoring, protection and black start. Local control and protection level encompasses primary voltage and frequency regulation, primary real and reactive power control for each local generation and energy storage unit.

To a large extent, the control of microgrids relies on information and communication technology (ICT). Therefore, it is necessary to discuss frequently-used technologies applied in distribution networks. The first of these are microprocessors, which are widely used in microgrids since they can provide support to make more complicated inverters and load controllers. With the rapid development of integrated circuit technology, future microprocessors are likely to become smaller, faster, cheaper and equipped with the ability to communicate. The second technology is communication. Communication networks offer adequate bandwidth to the users. In addition, the remote control of microgrids is highly reliant on good communication. The other two technologies are service oriented architectures (SOA) and the internet of energy. The former ensures the normal operation of microgrids in multiple layers and the latter uses software to remotely control home appliances through an internet gateway.

1.2.2 Centralized Control Versus Decentralized Control

Microgrids can either operate in centralized control mode or decentralized control mode [3]. In centralized mode, the Microgrid Central Controller (MGCC) plays the most important role in optimizing a microgrid. Based on electricity price, gas price and security information, the MGCC decides how much power is needed to be imported from the utility network and how many non-critical loads should be shed in critical conditions. In decentralized mode, the primary goal is to maximize power production to meet the load demands and export excess electricity to utility grid.

1.2.3 Forecasting

The implementation of both centralized control and decentralized control strategies requires load forecasting, actual renewable resources power output estimation and information about electricity prices. Forecasting load and renewable generation requires data collection and a weather forecast, which may increase the operation cost [4,5]. Therefore, the benefit of forecasting should be greater than the extra cost involved. Currently, forecasting is mainly focused on electricity price, load demand and PV generation aspects for large interconnected systems. A number of forecasting methods are available and implemented in utility operation. The simplest forecasting method (persistent method) is to predict a variable based on its current value.

1.2.4 State Estimation

In addition to forecasting, state estimation is another issue associated with microgrid control. Due to redundant measurements, state estimation at the transmission level can reduce the uncertainty. But at distribution level, decisions need to be made without the support of sufficient data at middle-voltage level or low voltage level. Therefore, the undetermined parameters of the model need to be estimated.

The estimation problem consists of a parameter estimation principle and a parameter estimation algorithm. The common principles used are least squares, maximum likelihood, minimum variance, minimum risk, etc. The algorithms used can be divided into iterative and recursive algorithms. The iterative algorithms include Newton's method, gradient method and Gaussian method, whereas the recursive algorithms include real-time online estimation, point-by-point data processing and parameter updating [6].

1.2.4.1 Least Squares Estimation

Suppose there is a set of unknown parameters, $\mathbf{x} = (x_1, x_2, \ldots, x_n)^T$. The aim is to estimate the value of $\mathbf{x}$ with some measurements containing noise, $\mathbf{y} = (y_1, y_2, \ldots, y_k)^T$. To find the best estimate, the simplest case can be considered in which each $\mathbf{y}$ is a linear function of $\mathbf{x}$, with some measurement noise $\mathbf{v}$. Thus:

$$\mathbf{y} = \mathbf{Hx} + \mathbf{v} \tag{1.1}$$

where $\mathbf{v} = (v_1, v_2, \ldots, v_k)^T$, and $\mathbf{H}$ is a $k \times n$ matrix. So Eq. (1.2) can be obtained as follows:

$$\begin{pmatrix} y_1 \\ \vdots \\ y_k \end{pmatrix} = \begin{pmatrix} H_{11} & \cdots & H_{1n} \\ \vdots & \ddots & \vdots \\ H_{k1} & \cdots & H_{kn} \end{pmatrix} \begin{pmatrix} x_1 \\ \vdots \\ x_n \end{pmatrix} + \begin{pmatrix} v_1 \\ \vdots \\ v_k \end{pmatrix} \tag{1.2}$$

Given an estimate $\hat{\mathbf{x}}$, the difference between the noisy measurements and the expected values is:

$$\boldsymbol{\varepsilon} = \mathbf{y} - \mathbf{H}\hat{\mathbf{x}} \tag{1.3}$$

Using the least squares principle, we will try to find the value of $\hat{\mathbf{x}}$ that minimizes the cost function:

$$J(\hat{\mathbf{x}}) = \boldsymbol{\varepsilon}^T\boldsymbol{\varepsilon} = (\mathbf{y} - \mathbf{H}\hat{\mathbf{x}})^T(\mathbf{y} - \mathbf{H}\hat{\mathbf{x}}) = \mathbf{y}^T\mathbf{y} - \hat{\mathbf{x}}^T\mathbf{H}^T\mathbf{y} - \mathbf{y}^T\mathbf{H}\hat{\mathbf{x}} + \hat{\mathbf{x}}^T\mathbf{H}^T\mathbf{H}\hat{\mathbf{x}} \tag{1.4}$$

The necessary condition for the minimization is that the partial derivative (gradient) of J with respect to $\hat{\mathbf{x}}$ equals zero.

$$\frac{\partial J}{\partial \hat{\mathbf{x}}} = -2\mathbf{y}^T\mathbf{H} + 2\hat{\mathbf{x}}^T\mathbf{H}^T\mathbf{H} = 0 \tag{1.5}$$

Then we get:

$$\hat{\mathbf{x}} = (\mathbf{H}^T\mathbf{H})^{-1}\mathbf{H}^T\mathbf{y} \tag{1.6}$$

The inverse $(\mathbf{H}^T\mathbf{H})^{-1}$ exists if $k > n$ and $\mathbf{H}$ is non-singular [7].

1.2.4.2 Weighted Least Squares Estimation

Suppose that confidence is not equal on all measurements. For example, some of parameters are measured with low noise, while others are measured with high noise. The other factors are the same as those described in Section 1.2.4.1, so the process can also start with function (1.1). Assume that each measurement may be taken under different conditions so that the variance of the measurement noise may be distinct too:

$$E\left(v_i^2\right) = \sigma_i^2, \quad 1 \le i \le k \tag{1.7}$$

Assume that the noise for each measurement has zero mean and is independent [8]. The covariance matrix for all measurement noise is:

$$\mathbf{R} = E(\mathbf{v}\mathbf{v}^T) = \begin{pmatrix} \sigma_1^2 & \cdots & 0 \\ \vdots & \ddots & \vdots \\ 0 & \cdots & \sigma_k^2 \end{pmatrix} \tag{1.8}$$

The sum of squared differences weighted over the variance of the measurements can be minimized.

$$\text{Min } J(\hat{\mathbf{x}}) = \boldsymbol{\varepsilon}^T\mathbf{R}^{-1}\boldsymbol{\varepsilon} = \frac{\varepsilon_1^2}{\sigma_1^2} + \frac{\varepsilon_2^2}{\sigma_2^2} + \cdots + \frac{\varepsilon_k^2}{\sigma_k^2} \tag{1.9}$$

Equation (1.9) can be expanded as:

$$J(\hat{x}) = \varepsilon^T R^{-1} \varepsilon = (y - H\hat{x})^T R^{-1} (y - H\hat{x})$$
$$= y^T R^{-1} y - \hat{x}^T H^T R^{-1} y - y^T R^{-1} H\hat{x} + \hat{x}^T H^T R^{-1} H\hat{x}$$

$$(1.10)$$

Therefore, the final estimation result is:

$$\hat{x} = (H^T R^{-1} H)^{-1} H^T R^{-1} y \qquad (1.11)$$

1.2.4.3 Newton-Raphson Algorithm

The Newton-Raphson algorithm is used to find the roots of a system of equations [9,10]. Assume there is an equation to be solved as follows:

$$f(x) = 0 \qquad (1.12)$$

Solving this requires two steps: (i) Select an initial value $x^{(0)}$ which is close to the zero point; (ii) draw a tangent line through the point $(x^{(0)}, f(x^{(0)}))$ and calculate the intersection point between the tangent line and x-axis. This point can be called $(x^{(1)}, 0)$.

$$x^{(1)} = x^{(0)} - \frac{f(x^{(0)})}{f'(x^{(0)})} \qquad (1.13)$$

$x^{(1)}$ should be closer to the zero point than $x^{(0)}$. Finally, after n iterations, the roots can be obtained.

$$x^{(n)} = x^{(n-1)} - \frac{f(x^{(n-1)})}{f'(x^{(n-1)})} \qquad (1.14)$$

When solving a system of nonlinear equations, the principle is the same. Assume there is a system of equations:

$$f(x) = 0 \qquad (1.15)$$

The system of equations has m equations and m unknown variables. So the equation can also be written as another form:

$$\begin{cases} f_1(x_1, x_2 \ldots x_m) = 0 \\ \quad \ldots \\ f_m(x_1, x_2 \ldots x_m) = 0 \end{cases} \qquad (1.16)$$

The roots of the system of equations are computed as follows:

$$\mathbf{x}^{(\mathbf{n})} = \mathbf{x}^{(\mathbf{n}-1)} - \mathbf{J}^{-1}\mathbf{f}(\mathbf{x}^{(\mathbf{n}-1)}) \tag{1.17}$$

$$\mathbf{x}^{(\mathbf{n})} = \left(x_1^{(n)}, x_2^{(n)} \dots x_m^{(n)}\right)^{\mathbf{T}} \tag{1.18}$$

$$\mathbf{x}^{(\mathbf{n}-1)} = \left(x_1^{(n-1)}, x_2^{(n-1)} \dots x_m^{(n-1)}\right)^{\mathbf{T}} \tag{1.19}$$

$$\mathbf{J} = \begin{pmatrix} \dfrac{\partial f_1(\mathbf{x}^{(\mathbf{n}-1)})}{\partial x_1^{(n-1)}} & \cdots & \dfrac{\partial f_1(\mathbf{x}^{(\mathbf{n}-1)})}{\partial x_m^{(n-1)}} \\ \vdots & \ddots & \vdots \\ \dfrac{\partial f_m(\mathbf{x}^{(\mathbf{n}-1)})}{\partial x_1^{(n-1)}} & \cdots & \dfrac{\partial f_m(\mathbf{x}^{(\mathbf{n}-1)})}{\partial x_m^{(n-1)}} \end{pmatrix} \tag{1.20}$$

1.3 MICROGRID CONTROL METHODS

In a microgrid, different kinds of control methods are applied to ensure reliable operation, in both grid-connected mode and islanded mode. Depending on the DG and operating conditions, there are three main types of control methods: PQ control, V/f control and droop control.

1.3.1 PQ Control

The main objective of PQ control is to keep the microsource's active power and reactive power constant when the frequency and voltage deviation stay within prescribed limits. In PQ control, the active and reactive power are firstly decoupled in order to achieve independent control. The active power controller aims to maintain the active power output constant at a given reference value within the permissible frequency range. The reactive power controller aims to maintain the reactive power output constant at the given reference value within the permissible voltage range. However, this PQ control method cannot maintain the frequency and voltage constant, so an extra distributed generator is needed to regulate the voltage and frequency of the microgrid within the acceptable range. If microgrid operates in the grid-connected mode, the main power grid is responsible for maintaining the voltage and frequency of the microgrid.

1.3.2 V/f Control

The main objective of V/f control is to maintain the system frequency and voltage magnitude constant regardless of the actual active and reactive power outputs of microsource. A frequency controller adjusts the active power output to maintain the frequency at the given reference value. A voltage controller adjusts the reactive power output to maintain the voltage at the given reference value. V/f control is common when the microgrid operates in islanded mode.

1.3.3 Droop Control

1.3.3.1 Active Power Control

In a microgrid, the load keeps changing all the time, so the generators will change their power output based on the frequency deviation. The relationship between active power output and frequency can be described by the following equation and Figure 1.3.

$$\Delta P = P_2 - P_1 = S_p(f_1 - f_2) \tag{1.21}$$

where ΔP = power output change of the generator; S_p = reciprocal of slope of curve, kW/Hz or MW/Hz, which is determined by characteristic of each DG.

1.3.3.2 Voltage Control

The linear relationship between reactive power and terminal voltage, as shown in Figure 1.4, is similar to that of active power and frequency. The system voltage control can be carried out by adjusting the reactive power output of microsources.

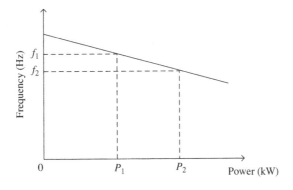

Figure 1.3 Relationship between active power output and frequency.

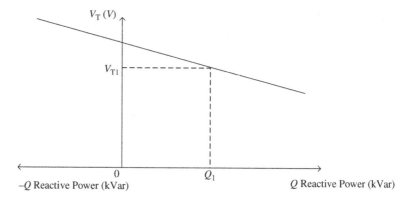

Figure 1.4 Relationship between reactive power output and voltage.

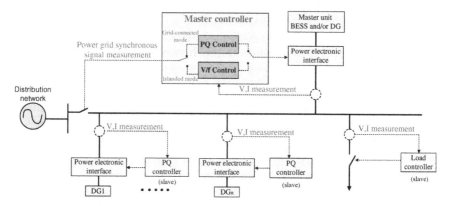

Figure 1.5 Master-slave control structure.

1.4 CONTROL ARCHITECTURES IN MICROGRIDS

Compared with a conventional power grid, microgrid has two different modes, grid-connected mode and islanded mode (grid-disconnected) mode. The microgrid should be able to operate reliably in both modes. Depending on the roles of distributed generation in the microgrid, the control architectures can be either master-slave control, peer-to-peer control or hierarchical control.

1.4.1 Master-Slave Control

Master-slave control structure is illustrated in Figure 1.5. When a microgrid operates in islanded mode, either the DG or ESS of the master unit will take V/f control role to provide voltage and frequency references for other DGs and ESS within the microgrid. Meanwhile, other DGs are in PQ control mode. The controller with V/f control method is named the master controller while the other controllers are slave controllers. Slave controllers take corresponding actions based on those of the master controller. However, microgrid operates in grid-connected mode in most situations. In such cases, the main grid provides voltage and frequency references for microgrid, so all the controllers within microgrid are in PQ control mode. If a fault occurs on the main grid side, the microgrid can seamlessly transfer from interconnected to islanded mode. One of DG's controls (master unit) needs to switch to V/f control mode.

The commonly-used master units can be divided into three categories: ESS, DG and ESS integrated with DG. If ESS is used as the master controller, the microgrid cannot operate in islanded mode for very long since ESS keeps discharging and will eventually run out of power. If a DG like a micro-turbine is used as the master controller, the voltage and frequency can be regulated easily, so the microgrid can operate in islanded mode for a long time. Another method is to use both DG and ESS as master controllers. This method works for renewable generators (PV, wind turbine, etc.). Due to the intermittency and stochastic nature of renewable generation outputs, ESS can help reduce the voltage and frequency fluctuation so that the microgrid is able to run in islanded mode for a long time.

If the energy capacity of the ESS is large enough, it can act as a master unit (V/f mode) to offer the specified voltage and frequency support for the islanded microgrid and adjust the power output to bring the frequency and voltage back to the scheduled values. In contrast, other DGs still remain as slave units operating in constant power (PQ) mode [11−13]. The ESS is required to mitigate the dynamic mismatch between load and renewable power generation in the islanded microgrid. However, it should be noted that the ESS cannot maintain at discharging state for a long time due to the State of Charge (SOC) limit, so islanded operation with only ESS as a master unit can only be sustained for a relatively short period of time.

If the energy capacity of ESS is small and limited, it can be integrated into the DC-link of a certain renewable power source as a combined master unit (V/f control mode) to provide the desired voltage and frequency support for an islanded microgrid [14]. Under these circumstances, a coordinated V/f and P/Q control strategy is required to realize the frequency regulation of the islanded microgrid [15]. Provided that the renewable power output from the PV or wind turbine generator (WTG) is larger than active power demanded for the frequency regulation of islanded microgrid and meanwhile state of charge (SOC) of battery ESS (BESS) is lower than upper limit (e.g., 80%), the battery will be charged to maintain the power balance between the generation and load to ensure frequency stability of the islanded microgrid. If the same situation occurs but SOC is higher than 80%, the battery cannot be charged due to this upper limit. In this situation, a central microgrid controller acts to smoothly change the operating state of this renewable power generator from V/f control mode to constant PQ mode, since BESS fails to assist in maintaining the desired power outputs for V/f control. At the same time, one DG with steady and sustained output (e.g., a micro-turbine generator) can operate in V/f mode. Similarly, when the renewable power output is smaller than the active power demanded for the frequency regulation of the islanded microgrid and also the SOC of BESS is higher than lower limit (e.g., 20%), the battery will be discharged to maintain the microgrid frequency. If the same scenario happens but SOC is less than 20%, the battery is unable to be discharged for frequency regulation and renewable power generator operation is switched to constant PQ mode. In this case, the V/f regulation task will be undertaken seamlessly by another DG featured with steady and sustained output in the microgrid. Using this method, the advantages of renewable energy sources can be fully utilized together with the rapid power response of the BESS so as to achieve sustained and steady islanded operation of the microgrid, in which frequency stability can be achieved. Compared with the application of ESS with large capacity, this control mode not only can minimize the demand on the ESS capacity, but also improve the overall operationing economics of the microgrid [16].

However, this control method has some disadvantages. First of all, it needs many communication channels between different controllers, which will increase the total investment cost of the microgrid. Secondly, it is difficult to apply the master-slave control method to larger systems. Lastly, implementation of the master-slave control method has strict requirements for communication and supervisory control.

1.4.2 Peer-to-peer Control

For peer-to-peer control of an islanded microgrid, all the DGs and ESS units are equal with no master-slave relation involved, as shown in Figure 1.6. Using the droop control method, designated DGs and

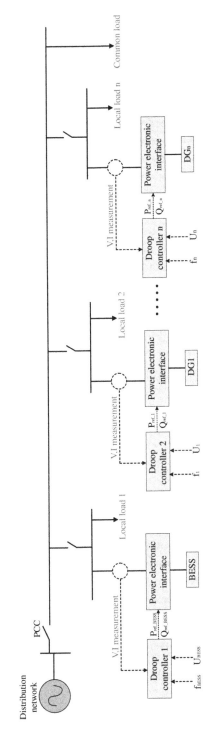

Figure 1.6 Peer-to-peer control structure.

ESS units are capable of adjusting their power outputs based on the locally measured voltage and frequency [17−18]. Any large change in load can be actively shared among selected DGs and ESS units based on individual droop coefficients so that power balance between generation and consumption is re-established in islanded operation. In a similar manner to the frequency droop control of a traditional synchronous generator, this peer-to-peer control belongs to error frequency regulation in that it allows some frequency and voltage deviation from the desired values. Compared with master-slave control, peer-to-peer control enables each DG and ESS to automatically participate in the power output allocation, which facilitates the plug and play function of DG [19]. Besides, the expense on communication system can be avoided so the system total cost is reduced accordingly. Meanwhile, there is no change in droop control strategy for DGs and BESS regardless of microgrid operation mode, so seamless switching between grid-connected and islanded mode can be easily achieved. In the practical microgrid application with the peer-to-peer control strategy used, some DGs can still adopt PQ control to realize the Maximum Power Point Tracking (MPPT) and unity power factor operation. Meanwhile, other DGs and ESS units rely on droop controls to undertake the power sharing task, which is otherwise fulfilled by the master unit in master-slave control. Through proper setting of the droop coefficients, the net power change can be shared among different DGs in order to achieve dynamic power balancing and maintain voltage and frequency within the acceptable range in islanded mode.

The control method used is droop control, as mentioned in Section 1.3.3. If frequency is decreased, the DG will increase its actual power output to maintain the frequency within acceptable limits. Otherwise, the active power output will be decreased. The same principle can be applied in the peer-to-peer voltage control of microgrids. If the voltage level drops, reactive power output will be increased accordingly. This process is described in Eq. (1.22).

$$\begin{cases} P = P_0 + (f_0 - f)K_f \\ Q = Q_0 + (U_0 - U)K_u \end{cases} \tag{1.22}$$

Finally, the whole system will operate at a new frequency and voltage.

If peer-to-peer control method is compared with the master-slave control method, it can be seen that the controllers in peer-to-peer control architectures can make decisions using local information, which means that this architecture can save a lot of money during establishing communication system and also minimize the system complexity. Another advantage is that these architectures can fulfill the requirements for seamless transition between grid-connected and grid-disconnected modes.

Currently, it is easy to apply peer-to-peer control method in a plug and play network, but it is still not widely implemented in practical application of microgrids.

1.4.3 Hierarchy Control

In hierarchical control, a central controller is implemented to send control signals to each DG, ESS unit and controllable loads. One type of two-layer control structure is illustrated in Figure 1.7 [20]. The first objective of the central controller is to predict the load demand and renewable power generation, so a set of operation plans is developed accordingly. Based on the collected status information including voltage, current and power, this operation plan can be updated in real-time to adjust the power output and determine the start and stop of DGs, loads and ESS units. In this way, the stability of voltage and frequency is ensured and relevant protection function is provided for the islanded microgrid as well. Regarding this hierarchical control scheme, physical communication channels are required for mutual communication between DGs and top-level controller. If one channel fails to work, the entire microgrid fails to operate normally.

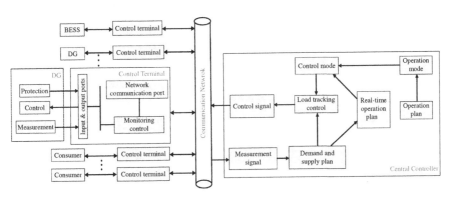

Figure 1.7 Typical two-layer structure.

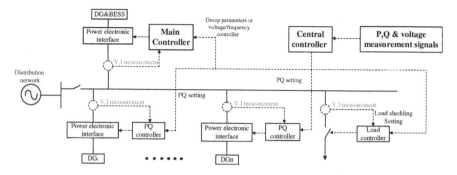

Figure 1.8 Two-layer control structure with weak communication connection.

In another type of hierarchical two-layer control structure, shown in Figure 1.8, only weak communication is needed to coordinate the central controller with local controllers of DGs [21]. Using this method, transient power demand-supply balance can be achieved by the low-level controllers of DGs, and meanwhile the top-level central controllers are capable of modifying the steady-state operating points of low-level DGs and managing the load based on the variation in both DG outputs and load demands. Even if communication fails for a short while, microgrid can still maintain normal operation during the period of fault.

A type of three-layer control structure is shown in Figure 1.9 [22]. The top-level operation and management system of the distribution grid (OMDG) is responsible for monitoring the real-time operation status of the microgrid group, which consists of multiple microgrids. The system manages and dispatches these microgrids in accordance with a power market mechanism and the dispatching command from OMDG. The middle-level MGCC is in charge of monitoring the operation status such as the key node's voltage and frequency of each microgrid in the islanded mode, power flow through each branch, current output and power reserve margins of each DG and ESS, load condition and so on. In addition, the MGCC can optimize the economic operation of each individual microgrid and provide ancillary services through the proper regulation of the ESS and DGs, including load following, operational reserve as well as frequency regulation and voltage control [14]. The bottom-level local controller consists of microsource controllers (MC) for the DG and the ESS as well as the load controller (LC), aiming to ensure a transient power

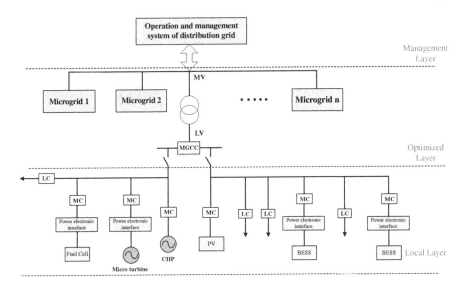

Figure 1.9 Three-layer control structure.

balance, power quality improvement for sensitive load and dynamic load management. The complete hierarchical control strategy can be realized through the multi-agent system (MAS), which comprises main-grid agents, microgrid agents and microsource agents [23,24]. Compared with the two types of control strategy described earlier in this section, a hierarchical control-based multi-agent method can enable a microgrid group to operate in an efficient, accurate and flexible manner. Not only can frequency regulation and voltage stability be achieved through the local controller and the MGCC, but also optimized economic operation with multiple objectives as well as coordinated control between microgrid and main grid can be realized.

From the discussion in this section, it can be concluded that stable microgrid operation cannot be achieved without intelligent control strategies. Choosing and designing the suitable control strategies is essential to achieve a safe and reliable microgrid operation.

1.5 MICROGRID PROTECTION

A microgrid protection scheme differs from that of conventional low-voltage or medium-voltage distribution network. It is still a main technical problem that needs to be tackled. The protection strategy needs to take into account faults in the main power grid and also faults inside the microgrid. If a fault occurs in the utility grid, protection devices in microgrid will trip as soon as possible to cut off the interconnection between the microgrid and its upstream grid to protect the loads within the microgrid. A commonly-used method is to combine circuit breakers with directional over-current relays or electronic static switches. If a fault occurs within the microgrid, protection devices will isolate the smallest possible area of distribution feeder to separate out the fault.

Conventional protection strategies use the magnitude and direction of the fault current to determine that a fault is occurring. Protection strategies should be reliable, secure, redundant and cost-effective. However, the inverter-interfaced DERs make system protection complex. In general, the principle associated with microgrid protection can be described as "3S": selectivity (whether to trip or not), sensitivity (whether a fault can be detected) and speed (the time for trip).

Several control and protection methods regarding microgrid are discussed in [25–28]. Some of the challenges of microgrid protection are:

1. Decide whether to connect or disconnect its upstream grid according to the magnitude of bi-directional short-circuit current.
2. Circuit breaker false trip due to the integration of DERs.
3. Contradictions between microgrid feeder protection and fault-ride-through requirement set by utility companies.
4. Interference caused by the power electronics components.

Unexpected relay tripping issues can be addressed by using adaptive protection, namely over-current protection relays combined with identification of current direction [29]. The definition of adaptive protection is "an online activity that modifies the preferred protective response to a change in system conditions or requirements in a timely manner by means of externally generated signals or control action" [30].

The design of modern adaptive protection schemes for microgrids can be divided into two types, namely pre-calculated setting group and real-time calculated setting group. Adaptive protection can also be classified as centralized or decentralized depending on control methods.

For an adaptive protection scheme with pre-calculated settings, the MGCC exchanges information with each circuit breaker and directional over-current relay through the RS-485 serial communication bus. When a fault happens, every relay makes its own decision by checking if the measurement value meets the pre-calculated tripping condition.

An adaptive protection scheme with real-time calculated settings is applied as a multifunctional intelligent digital relay (MIDR), whose structure is much more complicated than that of the adaptive protection scheme with pre-calculated settings. In general, the MIDR produces selective tripping signals and sends them to corresponding circuit breakers if a fault occurs. MIDRs can monitor both analog and digital signals.

Choosing the proper protection strategy is very important for maintaining the normal operation of a microgrid. The major problems of microgrid protection lie in variable operation conditions when renewable resources are integrated, lower sensitivity of fault current due to power electronic devices (like inverters) and false tripping. The solution to these problems is to apply an adaptive protection scheme that changes the protection settings automatically based on the configuration of microgrid. There is still great potential for development in this area. With more and more advanced protection devices developed and new relevant communication technologies applied, microgrids will become more reliable and popular in future power grids.

1.6 THREE-PHASE CIRCUIT FOR GRID-CONNECTED DG

The structure of a three-phase grid-connected DG system is shown in Figure 1.10; the system is composed of four parts: a DG, a three-phase inverter, an LC filter and an isolation transformer. Via the three-phase inverter and LC filter, the DG can generate AC power that meets the requirements of grid integration. Some of the power output is used to supply the loads and the remainder is delivered to the main grid at PCC via isolation transformer.

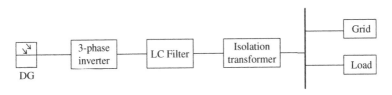

Figure 1.10 Three-phase grid-connected DG system structure.

1.6.1 LC Filter

The harmonics from inverters are mainly made up of the carrier frequency and its integer multiples. In order to avoid the resonance effect and make the filter output close to ideal sinusoidal waveform, the cutoff frequency of LC filter must be far lower than the frequency of the lowest-order harmonic of the pulse width modulation (PWM), and at the same time much greater than the fundamental frequency. So the cutoff frequency f_{Lc} can be chosen by using the formula in Eq. (1.23):

$$10 f_b < f_{Lc} = 1/2\pi\sqrt{LC} < f_s/10 \qquad (1.23)$$

In Eq. (1.23), f_b is the fundamental frequency; f_{Lc} is cutoff frequency of the LC filter; and f_s is the carrier frequency of sine pulse width modulation (SPWM).

The primary factor to be considered for an inductor is to minimize its impact on the voltage drop. The lower the inductance is, the smaller the output impedance of the filter. Commonly, the voltage drop is limited within 3% ~ 5%. Moreover, the effective value of harmonics should be no more than 10% ~ 20% of the inverter capacity, otherwise the inverter will enter its protection state.

There is a tradeoff between capacitance value selection and inductance value selection. According to the cutoff frequency formula, when a cutoff frequency is chosen, then the product of L and C is a constant. If the capacitance is too small, then the inductance would become very large, which would increase the voltage drop.

1.6.2 Isolation Transformer

The impact of DG on the distribution network (e.g., voltage, frequency, short circuit current) can be mitigated by installing an isolation transformer. Selecting the appropriate transformer parameters and determining the proper connection mode can achieve the following objectives:

1. Prevent the third harmonics from entering the utility grid.
2. Avoid DC current entering the utility grid.
3. Limit the magnitude of any short circuit current if a grid fault happens.

1.7 ENERGY STORAGE TECHNOLOGY IN RENEWABLE MICROGRIDS

This section discusses energy storage technology that facilitates high penetration and integration of variable renewable energy sources in a microgrid. ESS has been utilized in power systems for several decades in the United States, and now the integration of renewable energy sources is creating demands for increasing DES sources [31,32]. ESS plays an extremely important role in improving the operating capabilities of microgrids. ESS can be divided into mechanical, electrochemical, chemical, electrical and thermal systems, as shown in Figure 1.11. In the following, an overview is provided of the major types of energy storage applied in microgrids.

1.7.1 Batteries

1.7.1.1 Lead-Acid Batteries

The lead-acid battery has over 100 years of history since its invention and is the most widely used rechargeable battery. Approximately 70% of lead-acid batteries are used for vehicles, 21% for communications, and 4% for other applications [33]. The benefits of lead-acid batteries include low cost, high efficiency and good surge capability. They are an excellent choice for uninterruptible power supply and can also be used for spinning reserve applications [34].

A high voltage can be achieved by simply connecting lead-acid battery cells in series. Each lead-acid cell, with a 2 V voltage, is made from a spongy pure lead cathode, a lead dioxide anode as well as a 20−40% solution of sulfuric acid that acts as an electrolyte. When the

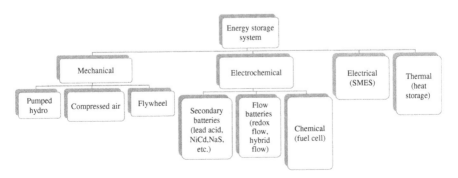

Figure 1.11 Classification of energy storage systems. Source: Fraunhofer ISE.

battery is discharging, a chemical reaction allows conduction between the anode and the cathode to generate electricity. The chemical reaction is reversible if the electrodes are supplied with voltage. That is the reason why lead-acid batteries can be recharged. The life cycle and the ability to tolerate deep discharges vary for different types.

No remarkable improvements have been achieved in the energy to volume ratio for a long time. Therefore, future battery research should pay more attention to other types of battery with more potential. Nevertheless, the lead-acid battery plays an important role today due to its cost-effectiveness. They are also easy to recycle. However, sometimes lead, which is a poisonous metal, can be a risk to the environment. Until more mature battery technologies appear, the lead-acid battery will still dominate the battery market.

1.7.1.2 Lithium-Ion Batteries

The light weight and quick response behavior make lithium suitable for making batteries. The power density of lithium-ion batteries is the best of all commercialized batteries. The voltage of a lithium-ion battery cell is 3.7 V, almost twice that of a lead-acid battery. The lithium-ion battery resembles a capacitor in the way it operates. There are three layers in lithium-ion battery. The first layer is the anode, and is made from a lithium compound. The second layer is the cathode and is made from graphite. The third layer is called the insulator, and it is located between the first and the second layer.

Lithium-ion has a very low self-discharge rate. In addition, the materials to make lithium-ion battery are abundant. The disadvantage is that lithium is expensive. Currently, cost and safety issues are the two main factors that impede the promotion of lithium-ion for widespread use in power systems.

Lithium-ion batteries are mainly used for small electronic devices like smartphones or laptops. For power systems, lead-acid batteries are more cost-effective at present, but lithium batteries have greater potential than other types of batteries. IBM is currently working on a project called Battery 500. The objective of this project is to develop a lithium battery that could store enough energy to power an electric vehicle for 500 miles.

1.7.1.3 Redox-Flow Battery

The electrolytes of redox-flow batteries can be exchanged. This feature makes it very easy to charge electric cars simply by refueling. However, the energy density of redox battery is very low, commonly 35 Wh/kg. But the Fraunhofer Institute in Germany claims that they could increase the redox-flow battery density up to the level of lithium-ion batteries (200 Wh/kg).

Other benefits of redox-flow battery include a long lifetime of about 40 years, and a capacity that can be easily increased by simply increasing the number of tanks and adding more electrolyte. With regard to grid storage, the capacity is very small, similar to that of flywheels, Superconducting Magnetic Energy Storage (SMES) or other battery storage types.

1.7.1.4 Sodium Battery

The sodium-sulfur battery is another type of battery under development. This technology has already been operated in some countries, such as Japan. The installed capacity of sodium battery is about 250 MW. The technological advantage of the sodium battery is its high power density, long battery lifetime (usually over 10 years) and high efficiency (up to 90%). But when this type of battery is operating, a high temperature (350°C) is required to liquefy the sodium. This constraint increases the difficulty and operational cost of sodium-sulfur batteries. Another drawback is that the sodium battery is very dangerous if the liquid sodium comes into contact with water in the atmosphere.

1.7.2 Flywheels

Flywheel energy storage unit (FESU) can supply immediate active power support for a renewable energy based microgrid. It has numerous merits such as high power density, high conversion efficiency and long life-span. In the past few decades, it has been used in uninterruptible power supplies where the short-duration power changes reduce the battery lifetime. In the context of autonomous energy production, flywheels are used in the field of transportation and in space applications for energy transfer and, particularly, to stabilize or drive satellites (gyroscopic effect) [35]. Flywheel energy storage is characterized by its long lifetime (typically 20 years) [36,37].

A flywheel is a disk with a certain amount of mass that can spin to store energy in kinetic form. To prevent the influence of gravity, the disk in flywheel ESS are built in perpendicular position of the rotor. Flywheels can be charged by electric motors when there is excessive electricity. It can also act as a generator when discharging.

Due to the existence of friction, eventually flywheels will lose some energy. Hence, minimizing friction can help to improve their efficiency. This goal can be realized through two approaches: the first one is to make a vacuum environment for the flywheel to spin in, ensuring there will be no air resistance. The second approach is to install a permanent magnet or electromagnetic bearing to make the spinning rotor float. The spinning speed of modern flywheel energy storage system can reach up to 16,000 rpm with a capacity of up to 25 kWh.

Flywheel have low maintenance costs, and their life-span can be long. There is no greenhouse emission or toxic material produced when flywheels are working, so it is very environment-friendly. The response time is very short. The drawbacks of flywheels are the small capacity and high power loss, ranging from 3% to 20% per hour.

1.7.3 Supercapacitor

The supercapacitor is developed from the electric double-layer theory. A two-layer capacitor is formed when the electrode is in touch with the electrolyte. Usually, the storage capacity of a supercapacitor is 20−1000 times that of the common capacitor. It has high power density and high energy conversion efficiency. Hence, it has attracted widespread attention. However, since the voltage of a single cell is low, a supercapacitor consists of numerous capacitors arranged in parallel or in series. Strictly speaking, the internal parameters of each capacitor are different and thus voltage imbalance may exist in supercapacitor and affect the operational reliability.

A practical supercapacitor for energy storage in microgrid requires a stack of many single cells connected in series [38]. Each cell consists of five layers with a porous separator in the center, a pair of porous electrodes on each side of it, and a pair of current collectors that congregate charges located at the end of each cell. The separator is an electrical insulator that prevents physical contact of electrodes but allows ion transfer between them [39−40]. The cells are packed and immersed into an electrolytic solution forming the double-layer charge distribution.

1.7.4 Comparison of Various ESS Technologies

Different types of energy storage technology are discussed in recent publications [41−44]. Characteristics of the different types of energy storage are compared in Tables 1.1 and 1.2. Large difference exists among different types of ESS. Thus, the best energy storage type should be chosen according to the practical microgrid application under consideration.

Different applications have different objectives and requirements, needing different features of each ESS technology. Therefore, it is

Table 1.1 Comparison of Characteristics for Different Types of Energy Storage

Type	Energy Density (Wh/kg)	Power Density (W/kg)	Response Time	Continuous Discharge Time	Cycling Times
Flywheel	5−30	400−1500	< 1 s	1 s−15 min	Above 20,000 times
Compressed air	30−60	–	1−10 min	Above 1−24 h	Above 100,000 times
Lead-acid	30−50	75−300	< 10 s	1 s−10 h	2000 times
Lithium-ion	75−200	150−300	< 10 s	1 s−10 h	< 10,000 times
Sodium-sulfur	100−250	100−230	< 10 s	1 s−10 h	2500−6000 times
Supercapacitor	5−10	5−10	< 1 s	1 ms−1 h	100,000 times
Superconducting magnetic	1−10	1−10	< 5 ms	1 ms−1 h	100,000 times

Table 1.2 The Advantages and Disadvantages of Different Types of Energy Storage

Type	Advantages	Disadvantages
Flywheel	High power density, quick response	Low energy density
Compressed air	Large capacity, long Continuous discharge time	Limited location, slow response
Lead-acid	Large capacity, high energy density	Low power density, low Cycling times
Lithium-ion	High energy density	Small capacity, hard for large scale application
Sodium-sulfur	High energy density	Needs high temperature in working process
Supercapacitor	High power density, quick response	Low energy density, expensive
Superconducting magnetic	High power density, quick response, high Cycling times	Low energy density, expensive, magnetic pollution to environment

necessary to conduct a comprehensive comparison and assessment of all storage technologies. When choosing an ESS for a microgrid application, the best balance of energy density, power density, response time and lifetime needs to be chosen.

ESS technologies can be categorized depending on discharge time and energy-to-power ratio:

- Short discharge time (seconds to minutes) is suitable for those applications whose energy to power ratio is less than 1: capacitor banks, superconducting magnetic energy storage and flywheels.
- Medium discharge time (minutes to hours) is suitable for those applications whose energy to power ratio is between 1 and 10: flywheel energy storage, lead-acid batteries, lithium-ion batteries and sodium sulfur (NaS) batteries.
- Long discharge time (days to months) is suitable for those applications whose energy-to-power ratio is greater than 10: hydrogen (H_2) and synthetic natural gas (SNG), pumped hydro storage, CAES, redox batteries.

1.7.5 Battery Energy Storage Modeling Consideration

The electrochemical model and equivalent electric circuit model are the common models for battery dynamic charging and discharging modeling. For the lead-acid battery, the simple equivalent circuit model, Thevenin equivalent circuit model, third order dynamical model and fourth order dynamical model are commonly used [45]. The fourth-order dynamical model proposed by Giglioli is a detailed model, but lots of experimental parameters are needed and the computation is complicated [46]. Figure 1.12 shows a typical constant-current discharging voltage characteristic curve of the nonlinear battery model proposed in [47].

As shown in Figure 1.12, the constant-current discharge voltage characteristic curve of battery can be divided into several parts, including the index-number characteristic part at the start of discharging and the rated characteristic part when the voltage change is small. The equivalent circuit model can be obtained by fitting the characteristic curve in Figure 1.12, as shown in Figure 1.13. The model has internal resistance R and a controlled voltage source E in series [47]. E and R are nonlinear functions of the SOC and temperature. Also, the internal

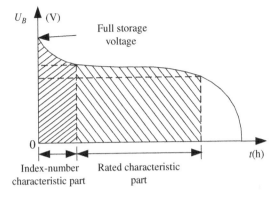

Figure 1.12 Typical constant-current discharging voltage characteristic curve.

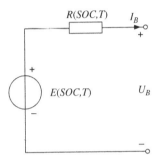

Figure 1.13 The equivalent circuit model of battery.

resistance takes a slightly different value during the charging and the discharging period [48].

Battery energy storage system usually has a short lifetime and high cost, and thus its lifetime should be a significant factor for the control and optimization of microgrids. ESS lifetime varies greatly depending on its usage [49]. Hence, a detailed battery model is required for accurate lifetime estimation.

A physical battery model with aging effect estimation considering SOC and state of health (SOH) is presented in [50]. For this method, both the SOC and SOH of the battery are calculated and updated for the optimization. SOH is obtained by considering two major factors, i.e., the cyclation based state of health (SOHc) and the calendrical aging (SOHt) [49]. An example Lead–Acid Battery Life loss cost model is introduced in [51].

REFERENCES

[1] Smart grids, European Technology Platform on Vision and Strategy for Europe's Electricity Networks of the Future 2006.

[2] Morris GY, Abbey C, Wong S, Joos G. Evaluation of the costs and benefits of microgrids with consideration of services beyond energy supply. In: IEEE PES General Meeting, 2012.

[3] Hatziargyriou N. Microgrids architectures and control. John Wiley & Sons; 2014.

[4] Charytoniuk W, Chan M. Short-term load forecasting using artificial neural networks. IEEE Trans Power Syst 2000.

[5] Hippert S, Pedreira C, Souza C. Neural networks for short-term load forecasting. IEEE Trans Power Syst 2001.

[6] Paleologu C, Benesty J, Ciochina S. A robust variable forgetting factor recursive least-squares algorithm for system identification. IEEE Signal Process Lett 2008;15:597–600.

[7] Simon D. Optimal state estimation. Hoboken (NJ): John Wiley & Sons, Inc; 2006.

[8] Bruno C, Candia C, Franchi L, Giannuzzi G, Pozzi M, Zaottini R, et al. Possibility of enhancing classical weighted least squares state estimation with linear PMU measurements. IEEE Bucharest Power Tech Conf 2009.

[9] Ben-Israel A. A Newton-Raphson method for the solution of systems of equations. J Math Anal Appl 1966;15(2):243–52.

[10] Ng SW, Lee YS. Variable dimension Newton-Raphson method. IEEE Trans Circuits Syst 2000;47(6):809–17.

[11] Bronsbergen CS. The first microgrid in the Netherlands. In: Proceedings of the Kythnos 2008 symposium on microgrids. Kythnos Island; 2008.

[12] Georgakis D, Papathanassiou S, Hatziargyriou N. Operation of a prototype microgrid system based on micro-sources equipped with fast-acting power electronics interfaces. In: Proceedings of the 35th power electronics specialists conference. Aachen; 2004. p. 2521–26.

[13] Hara R. Demonstration project of 5 MW PV generator system at Wakkanai. In: Proceedings of the Kythnos 2008 symposium on microgrids. Kythnos Island; 2008.

[14] Tan X, Li Q, Wang H. Advances and trends of energy storage technology in Microgrid. Int J Electr Power Energy Syst 2013;44(1):179–91.

[15] Adhikari S, Li F. Coordinated V-f and P-Q control of solar photovoltaic generators with MPPT and battery storage in microgrids. IEEE Trans Smart Grid 2014;5(3):1270–81.

[16] Guo L, Wang C. Dynamical simulation on microgrid with different types of distributed generations. Autom Electr Power Syst 2009;33(2):82–6.

[17] Wang C, Xiao Z, Wang S. Synthetical control and analysis of microgrid. Autom Electr Power Syst 2008;32(7):98–103.

[18] Chandorkar MC, Divan DM, Adapa R. Control of parallel connected inverters in stand-alone AC supply systems. IEEE Trans Ind Appl 1993;29(1):136–43.

[19] Lasseter R, Paigi P. Microgrid: a conceptual solution. In: Proceedings of power electronics specialists conference. Aachen; 2004. p. 4285–90.

[20] Funabashi T, Fujita G, Koyanagi K, et al. Field tests of a microgrid control system. In: Proceedings of the 41st international universities power engineering conference. Newcastle; 2006. p. 232–36.

[21] Pecas Lopes JA, Moreir ACL, Resende FO. Control strategies of microgrids black start and islanded operation. Int J Distri Energy Resour 2005;1(3):241–61.

[22] Dimeas AL, Hatziargyriou ND. A MAS architecture for microgrids control. In: Proceedings of thr 3th international conference on intelligent systems application to power systems. Washington; 2005. p. 402−06.

[23] Dimeas AL, Hatziargyriou ND. Operation of a multi-agent system for microgrid control. IEEE Trans Power Syst 2005;20(3):1447−55.

[24] Tsikalakis AG, Hatziargyriou ND. Centralized control for optimizing microgrids operation. IEEE Trans Energy Convers 2008;23(1):241−8.

[25] Al-Nasseri H, Redfern MA, O'Gorman R. Protecting micogrid systems containing solid-state converter generation. In: International conference on future power system. 2005.

[26] Vilathgamuwa DM, Loh PC, Li Y. Protection of microgrids during utility voltage sags. In: IEEE Transactions on Industry Electronics. 2006.

[27] Driesen J, Vermeyen P, Belmans R. Protection issues in microgrids with multiple distributed generation units. In: 4th Power conversion conference. 2007.

[28] Nukkhajoei H, Lasseter RH. Microgrid protection. In: IEEE PES general meeting. 2007.

[29] Hatziargyriou N. Microgrids architectures and controls. John Wiley & Sons; 2014.

[30] Rockefeller GD, et al. Adaptive transmission relaying concepts for improved performance. In: IEEE transaction on power delivery. 1998.

[31] Vartanian C. The coming convergence, renewables, smart grid and storage. In: IEEE energy 2030, November 2008.

[32] Hoffman MG, Sadovsky A, Kintner-Meyer MC, DeSteese JG. Analysis tools for sizing and placement of energy storage in grid applications, a literature review; 2010. Available from: <http://www.pnl.gov/main/publications/external/technical_reports/PNNL-19703.pdf>.

[33] Chang Y, Mao X, Zhao Y, Feng S, Chen H, Finlow D. Lead-acid battery use in the development of renewable energy systems in China. J Power Sources 2009;191(1):176−83.

[34] Chen H, Cong Y, Yang W, Tan C, Li Y, Ding Y. Progress in electrical energy storage system: a critical review. Prog Nat Sci 2009;19(3):291−312.

[35] Kascak PE, Kenny BH, Dever TP, Santiago W. International space station bus regulation with NASA Glenn research center flywheel energy storage system development unit. In: Proc. Intersoc. Energy Convers. Conf., Savannah, GA, July 29−August 2, 2001. p. 1−2. NASA/TM-2001-211138.

[36] Hebner R, Beno J, Walls A. Flywheel batteries come around again. IEEE Spectr 2002;39 (4):46−7.

[37] Rudell AJ. Flywheel energy storage renewable energy systems. Chilton, UK: CCLRC Rutherford Appleton Lab.

[38] Niu R, Yang H. Modeling and identification of electric double layer supercapacitors. In: Proc. of 2011 international conference on robotics and automation (ICRA 2011), 2011. p. 1−4.

[39] LeBlanc OH. Mathematics of ultracapacitors, GE Global Research, Technical Report, June 1993.

[40] Belhachemi F, Rael S, Davat B. A physical based model of power electric double-layer supercapacitors. Ind Appl Conf 2000;5:3069−76.

[41] Parker CD. Lead-acid battery energy storage systems for electricity supply networks. J Power Sources 2001;100:18−28.

[42] De Vries T, McDowall J, Umbricht N, Linhofer G. A solution for stability. Power Eng Int 2003;11(10):57−61.

[43] Tachibana Y. A new power supply system using NaS battery. In: Proc. 17th WEC Congr., 1998.

[44] Emura K. Recent progress in VRB battery. In: Proc. EESAT Conf, 2003.

[45] Salameh ZM, Casacca MA, Lynch WA. A mathematical model for lead-acid batteries[J]. IEEE Trans Energy Conversions 1992;7(1):93−7.

[46] Chan HL, Sutanto D. A new battery model for use with battery energy storage systems and electric vehicles power systems[C]. Power Eng Soc Winter Meeting 2000;1:470−5.

[47] Tremblay O, Dessaint L-A, Dekkiche A-I. A generic battery model for the dynamic simula-tion of hybrid electric vehicles[C]. In: Vehicle power and propulsion conference, Canada, September, 9−12, 2007. p. 284−89.

[48] Mi C, Masrur A, Gao W. Hybrid electric vehicles: principles and applications with practical perspectives. John Wiley & Sons; 2011.

[49] Su WF, Lin CE, Huang SJ. Economic analysis for demand-side hybrid photovoltaic and battery energy storage system. In: Proc. ind. appl. conf. 34th IAS annual meeting, vol. 3, 1999. p. 2051−57.

[50] Ye Y, Hui L, Aichhorn A, Jianping Z, Greenleaf M. Sizing strategy of distributed battery storage system with high penetration of photovoltaic for voltage regulation and peak load shaving. IEEE Trans Smart Grid 2014;5(2):982−91.

[51] Bo Z, Xuesong Z, Jian C, Caisheng W, Li G. Operation optimization of standalone micro-grids considering lifetime characteristics of battery energy storage system. IEEE Trans Sustainable Energy 2013;4(4).

CHAPTER 2

Applications of ESS in Renewable Energy Microgrids

2.1 INTRODUCTION

The operation and control of renewable microgrids is challenging. Due to the variability and intermittency of renewable distributed generators (DGs), it is difficult to keep the microgrid operating smoothly all the time. The weather-dependent DGs may not be able to accommodate the constantly varying load very well. Also, the fluctuation of renewable distributed generation may cause instability in the microgrid. Thus, energy storage systems (ESS) are very important in renewable microgrids, especially for islanded microgrids. Since an ESS can operate as load or generator when it is in charging or discharging mode, it is able to balance the power in the microgrid and reduce the impact of any fluctuation. This improves the stability of the microgrid significantly.

An ESS can be added in many different places in the microgrid depending on its purpose. When acting as a load, the ESS can perform load leveling and peak shifting functions to flatten the load variation and support the peak demand. For renewable generators, the ESS can help suppress any fluctuations. Also, the ESS is able to improve the power quality of the microgrid by working as an Uninterruptible Power System (UPS) and enhance the low voltage ride through (LVRT) ability for wind power and solar photovoltaic generation.

For different applications, the configuration of ESS may need to be different. There are two different configurations of ESS in a renewable microgrid—the aggregated ESS and distributed ESS. The different structure determines their performance and applications.

This chapter is organized as follows: in Section 2.2 and 2.3, two different configurations of ESS in the renewable microgrid will be introduced, namely aggregated and distributed ESSs. Then, Section 2.4 discusses energy management methods for handling load variation. In Section 2.5, the problem of generation fluctuation of renewables will

Energy Storage for Sustainable Microgrid. DOI: http://dx.doi.org/10.1016/B978-0-12-803374-6.00002-0

be presented, and three methods are presented to mitigate the fluctuation. The application of ESS as Uninterruptible Power System (UPS) will be introduced in Section 2.6. After that, in Section 2.7, the method to apply ESS for LVRT for wind power will be discussed. Then, power quality enhancement using ESS will be considered in Section 2.8. Using ESS to support voltage regulation is introduced in Section 2.9. ESS as spinning reserve will be presented in Section 2.10, followed by a case study in Section 2.11.

2.2 AGGREGATED ESS

In many microgrid projects, ESS is employed in an aggregated configuration. An aggregated ESS is a big energy storage facility with dedicated housing in the microgrid. It usually has a large capacity to store a huge amount of energy and high power output ability. An aggregated ESS may consist of several storage units. For example, a battery energy storage system (BESS) is composed of hundreds of battery packages, or a flywheel energy storage system (FESS) consists of many flywheel units.

Figure 2.1 shows a typical configuration of an aggregated ESS in a renewable microgrid. There is a photovoltaic (PV) farm in the microgrid as an aggregated renewable power plant. The aggregated ESS is connected on the same bus as the PV farm. So, the renewable power plant works with the aggregated ESS to provide stable power to the loads.

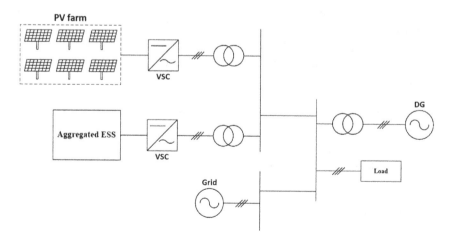

Figure 2.1 Aggregated ESS.

According to some research [1,2], an aggregated ESS is more capable of suppressing the power fluctuation in the microgrid than a distributed ESS. This is because the power output of the entire renewable farm is usually more stable than each individual unit. Also, the aggregated ESS has more capacity to stabilize the renewable farm since each distributed ESS only works for its own DG.

Besides the fluctuation suppression feature, aggregated ESSs also perform some energy management functions, such as load leveling and peak shifting. Moreover, in a PV microgrid, the aggregated ESS can store energy in the daytime and supply the microgrid in the night.

2.3 DISTRIBUTED ESS

There are two different configurations of distributed ESS. One is the distributed ESS on generator side. This helps the distributed generators to output smooth power. Another type is the distributed ESS on load side. This can be applied to reduce the load variation and manage energy for different purposes.

2.3.1 Generator Side Distributed ESS

An individual distributed ESS is smaller than an aggregated ESS, because it only handles a single (or a small group) renewable generation unit. Similar to aggregated ESSs, the major function of generator side distributed ESS is to smooth the output of renewables. The distributed ESSs are installed on-site with each renewable generation unit, as illustrated in Figure 2.2.

A distributed ESS is usually connected to the DC link of the renewable generation unit behind the grid-side inverter. For a wind inverter, the ESS connected to the DC link in the back-to-back converter. For solar PV generation, the ESS is connected to the output of PV through the DC/DC converter. In Figure 2.2, the distributed ESS is able to help the wind turbine inverter to have a stable DC link voltage, so the inverter can work properly. The distributed ESS outputs the desired power to compensate the fluctuation of renewable generation.

As mentioned before, a distributed ESS is not as powerful as an aggregated ESS for mitigating the fluctuation of renewables, but the distributed ESS is easier to be expanded and maintained.

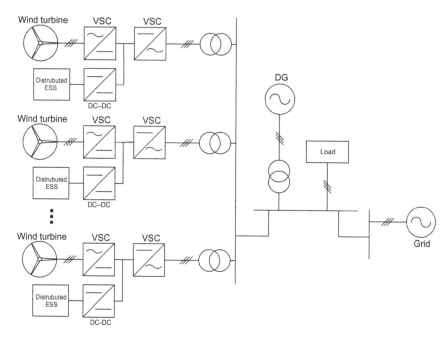

Figure 2.2 Generator side distributed ESS.

2.3.2 Load Side Distributed ESS

Another type of distributed ESS is the load-side distributed ESS. As Figure 2.3 shows, these are connected to the local loads.

A microgrid is a small power system that is not as strong as the main grid, so any variation in load may have bigger impact on the operation of the microgrid system. For example, when the load changes rapidly, it is necessary for the DGs to follow the load change. If the DGs cannot match the loads instantly, unstable operation may result. Also, load following by DGs is not energy efficient since the DGs cannot always work at maximum power point (MPP) or at the rated condition. Therefore, load leveling technology through distributed ESS can be used to smooth the load variation.

For the grid-connected microgrid, the load side distributed ESS can help users to save money. For example, the controller of distributed ESS can obtain the electricity price information from the power market. So, the distributed ESS can be charged from the main grid during the low electricity price period, and supply the loads from the stored energy when the price is high.

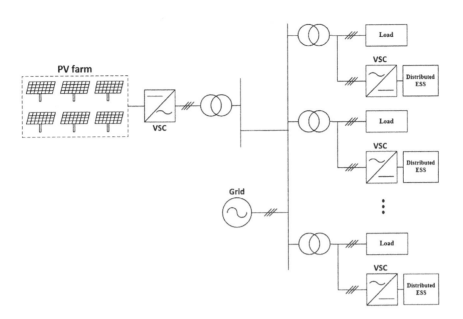

Figure 2.3 Load side distributed ESS.

Nowadays, as electric vehicles (EVs) become more popular, the possibility of using batteries on board EVs as the load side distributed ESS has been discussed in some publications. The related technologies are called Vehicle-to-grid (V2G) or Vehicle-to-home (V2H). The V2G is a technology that can use the stored electricity energy in an EV to help maintain the stability of the electrical grid. It is envisioned that hundreds of EVs will plug into microgrids in the future. The batteries in these EVs can be applied to help with the regulation of the grid. V2H is a technology to use EV batteries as an ESS for a house when the car is in the garage. The EV battery can be controlled to perform peak shifting of household loads to lower the electricity bill. Also, since the huge battery capacity of EVs can supply an average home's daily electricity demand for several days [3], EVs can also function as emergency power supply for households.

By implementing V2G and V2H, vehicle batteries need to be charged and discharged more frequently than in normal use, which may cause faster degradation of batteries. So V2G and V2H should be equipped with optimal strategy for battery charging and discharging by considering life cycle, economy and regulation performance.

2.4 ENERGY MANAGEMENT (LOAD LEVELING AND PEAK SHIFTING)

An ESS is useful for managing energy and improving the stability and economy of a microgrid system. The major energy management functions include load leveling and peak shifting. They are widely used to mitigate load fluctuations and improve power quality.

Load leveling and peak shifting are similar. Both of them deal with the variation of demand. However, the load leveling is more focused on short-term fluctuation, while peak shifting is more for the long-term (24 h) variation.

Figure 2.4 shows an example load profile. From this figure, we can see that the demand varies a lot during a day. The peak demand happens around 19:00, and may last about 2 h. But in the morning before 8:00, the load is very light. So it is possible for ESS to store energy in the early morning and output the energy to supply the microgrid in the early evening. This is called peak shifting.

Also, an example of short-term variation of loads is shown in Figure 2.5. Load leveling can be used to smooth this demand. As a result, the efficiency and power quality of the microgrid can be improved.

2.4.1 Load Leveling

Load leveling is an important function of the ESS. It is useful to reduce the influence of the load variation and lower the cost of the microgrid. The principle of load leveling is shown in Figure 2.6. The original load in the figure is changing constantly due to the switching on and off of the

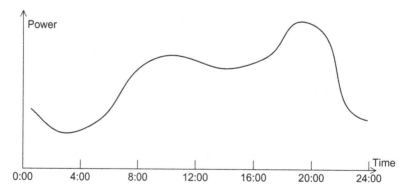

Figure 2.4 An example of daily demand profile.

devices in the system. With the help of the ESS, the sudden increase and decrease of load can be compensated for, which means that the ESS works like a floating load. The resulting load curve can keep constant for a duration if the mean value of the load does not change too much.

For an islanded microgrid, since there is no strong power source from a main utility grid, the load variation may cause the generation references of DGs to change dramatically. As a result, some DGs in the microgrid cannot work at their rated power output. So the overall microgrid is not efficient enough when the load is varying, and this may be harmful for the generators. To solve this problem, load leveling feature of ESS can be employed. ESS in the microgrid can store energy when the load is low, and output the energy when the load is high. Hence the variation of load can be compensated by the ESS.

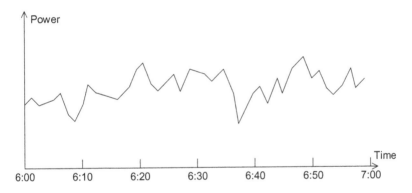

Figure 2.5 Demand variation within 60 min.

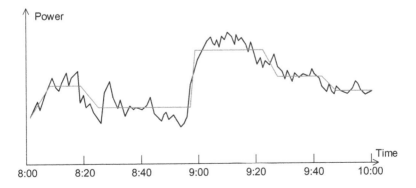

Figure 2.6 The principle of load leveling.

In the load leveling technique, it is important to know how much energy the ESS should output and how much it can store. To illustrate how to calculate the power reference for ESS, a microgrid with N DGs are considered here. The power balance equation (without considering the line impedance) can be denoted by (2.1).

$$P_D = \sum_{i=1}^{N} P_{DGi} + P_{ESS} \qquad (2.1)$$

where P_D is the real power demand of the loads; P_{DGi} is the real power generated by each DG in the microgrid; P_{ESS} is the output of the ESS. If $P_{ESS} > 0$, it means that the ESS is discharging. On the other hand, $P_{ESS} < 0$ means that the ESS is being charged.

For economical operation, it is better to keep the DGs working under constant output condition so that the generators can have high efficiency. Therefore, it is possible to assume that the total power generation $P_{totalG} = \sum_{i=1}^{N} P_{DGi}$ in the microgrid from all DGs is a fixed value. Then the output reference of the ESS can be obtained as (2.2).

$$P_{ESSref} = P_D - P_{totalG} \qquad (2.2)$$

And the ESS charging and discharging condition is as follows:

$$\begin{cases} \text{charging,} & P_D < P_{totalG} \\ \text{discharging,} & P_D > P_{totalG} \end{cases}$$

Also, the ESS can replace some DGs in the microgrid to supply the load during off-peak periods. So those less economical DGs (e.g., diesel generators) can be turned off to save money or fuel. In such situations, some DGs in the microgrid are turned off, so the total power generation P_{totalG} can be reduced to $P'_{DG} = \sum_{i \in M} P_{DGi}$ and $M = \{i | i \in (1, N) \text{ and DG}i \text{ is working}\}$ indicates the set of operating DGs. So the output reference of ESS becomes (2.3).

$$P_{ESSref} = P_D - P'_{DG} \qquad (2.3)$$

It is worth noting that the maximum output ability of ESS should satisfy the biggest difference between the load and generation, which is $P_{ESS,max} > P_{D,max} - P_{totalG}$.

2.4.2 Peak Shifting

In a microgrid, the ESS can function either as a load (during charging period) or as a generator (when discharging). Therefore, it is able to perform peak shifting function to reduce or eliminate the peaks and

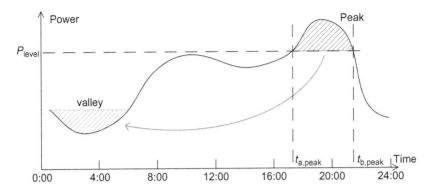

Figure 2.7 The principle of peak shifting.

valleys in the load profile, so the microgrid can satisfy the high demand during peak time. Also, this is useful to improve the economy of the microgrid operation. The principle of peak shifting is illustrated in Figure 2.7.

Typically, peak shifting is based on load forecasting technology. From the load forecasting results, the microgrid is provided with the information of how much energy or power is needed during peak times. Then, the ESS in the microgrid can store the required energy during valley period and support the system in the peak time. Nowadays, load forecasting is very precise. For some regions, the prediction errors can be as low as $1-3\%$ [4].

Suppose that during the peak time, the total power output ability of the DGs in the microgrid is $P_{DG,peak}$.

$$P_{DG,peak} = \sum_{i \in M_{peak}} P_{DGi} \qquad (2.4)$$

where M_{peak} is the set of all DGs that will operate during the peak period. The desired load level P_{level} is set to be lower than the total power output $P_{DG,peak}$ so that the generation can meet the demand in the peak level. The duty of the ESS is to reduce the load to the desired P_{level}. On the other hand, since the next day's load can be forecasted with good accuracy, the predicted demand can be denoted by $P_D(t)$. So the required energy for the ESS during the peak time $(t_{a,peak}, t_{b,peak})$ can be calculated from Eq. (2.5).

$$E_{peak} = \int_{t_{a,peak}}^{t_{b,peak}} (P_D(t) - P_{level}) \, dt \qquad (2.5)$$

Now, as the total energy needed for ESS during the peak time is known, the goal is to make sure that the ESS stores enough energy before the peak period. So the problem can be converted to finding how much state of charge (SOC) the ESS should reserve before the peak time. Assuming that the rated capacity of the ESS is E_{rated}, the reserved SOC is SOC_{res}, and SOC_{min} denotes the minimum SOC for the ESS. Then, the following relationship can be obtained.

$$(SOC_{res} - SOC_{min}) \cdot E_{rated} = E_{peak} \qquad (2.6)$$

Then, the reserved SOC should be

$$SOC_{res} = \frac{E_{peak}}{E_{rated}} + SOC_{min} \qquad (2.7)$$

Therefore, the ESS should be charged during the off-peak time, and its SOC should be above SOC_{res} at the beginning of peak time $t_{a,peak}$.

Note that $SOC_{res} < SOC_{max}$ should be satisfied, otherwise the ESS will be overcharged. Also, the desired load level P_{level} should meet the following constraint:

$$P_{D,max} - P_{level} < P_{ESS,max} \qquad (2.8)$$

where $P_{D,max}$ is the maximum demand power and $P_{ESS,max}$ is the maximum output ability of the ESS.

For a grid-connected microgrid, the peak shifting can also help to reduce the cost of microgrid by reducing power imported from utility grid during peak time.

2.5 FLUCTUATION SUPPRESSION (INTERMITTENCY MITIGATION)

In many microgrids, renewables are the major generation source, especially during islanded operation. Renewables, like wind and solar, are usually unsteady power sources. The output of wind turbines and solar panels are intermittent due to weather variations such as clouds over the photovoltaic arrays and other factors such as the wind turbine wake effect and the tower shadow effect. In addition, the microgrid is not as strong as the utility grid, so the fluctuation of generation may cause remarkable variations in network frequency and voltage, which makes the microgrid unstable. To solve this problem, ESS can be employed to mitigate the fluctuation.

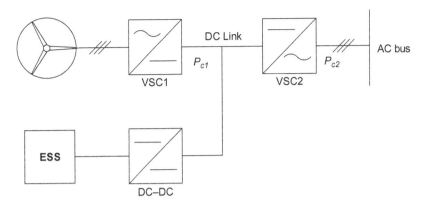

Figure 2.8 Wind turbine microgrid system with ESS.

In this section, three different fluctuation suppression techniques will be introduced. They are constant power control, output filtering and ramp-rate control. The objective of constant power control is to keep the output of renewable generation constant. Output filtering aims to utilize the ESS to simulate a low-pass filter so that the high frequency terms can be eliminated. Finally, ramp-rate control aims to keep the change rate of renewable power generation within a desired range.

2.5.1 Constant Power Control

Constant power control is a simple way to reduce the output fluctuation of renewable generation. In a microgrid, it is best to keep the output of renewable DG at constant for a given period of time. This will make renewables to be reliable power sources in microgrid. As an example, a wind turbine system with an ESS is illustrated in Figure 2.8. The output of the wind generator is connected to a back-to-back converter [5], which consists of two voltage source converters (VSC). As shown, the DC link is located between the VSC1 and VSC2. The ESS's DC-DC converter is connected to the DC link of the back-to-back converter.

The control objective of the ESS is to regulate the output active power P_{c2} of VSC2, which is the interface with the main grid. According to Figure 2.8, the power balance equation of the system is (2.9).

$$P_{c2} = P_{ESS} + P_{c1} \qquad (2.9)$$

where P_{c1} is the output of VSC1 and P_{ESS} is the output of ESS. Now, assuming that the desired output of this wind power system is P_{c2}^*. According to the value of P_{c1} and P_{c2}^*, the operation mode of ESS can

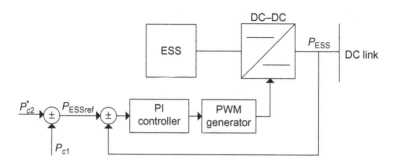

Figure 2.9 The constant control for ESS.

be determined: If $P_{c1} > P_{c2}^*$, the DC–DC converter of ESS works in buck mode and the ESS starts to be charged from the back-to-back converter. If $P_{c1} < P_{c2}^*$, the DC–DC converter of ESS works in boost mode, and the ESS is discharging to supply the microgrid.

By setting P_{c2}^* as a constant value, the microgrid output power can be fixed. Therefore, the reference of ESS is calculated by (2.10).

$$P_{ESSref} = P_{c2}^* - P_{c1} \qquad (2.10)$$

The PI controller can be utilized to control the DC–DC converter to follow the reference P_{ESSref}. The structure of the controller is given in Figure 2.9.

2.5.2 Output Filtering

Output filtering is a technology to let ESS act like a low-pass filter so that the high frequency terms of the renewable's output can be reduced, which makes the renewable's output smooth. Because the ESS needs to counteract the high frequency terms, the dynamic of ESS should be fast. Batteries and supercapacitors are fast enough to be employed in this technology. Figure 2.10 shows an example of PV microgrid with BESS [6].

Like the wind turbines, the output of solar panels is intermittent due to cloud cover. Therefore, an ESS is required to mitigate the fluctuation of the solar power output. Suppose that the output power of PV system is P_{pv}. Because the generation of solar panel may vary when they are shaded by clouds, the output P_{pv} of PV is stochastic. In order to stabilize the output, a low-pass filter can be utilized to obtain a smooth generation. The filtered power P_{fil} can be calculated by using (2.11).

$$P_{fil} = P_{pv} f(s) \qquad (2.11)$$

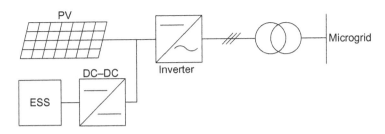

Figure 2.10 PV microgrid with ESS.

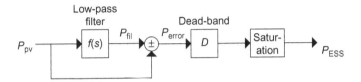

Figure 2.11 Control strategy to mitigate the PV output fluctuation.

where $f(s)$ is a low-pass filter. In a PV/ESS microgrid, the ESS can take charge to be the filter. The objective of the ESS is to compensate for the difference between the actual output and the filtered value (desired output). Hence, smooth output for the PV system can be achieved. The error between the actual output and the desired output can be obtained by (2.12).

$$P_{error} = P_{fil} - P_{pv} = P_{pv}f(s) - P_{pv} \qquad (2.12)$$

Theoretically, this error can be the output reference of the ESS. However, since the output of PV is changing frequently, the ESS will constantly be in charging or discharging mode. This may be damaging for the health of the ESS, especially for batteries, which have a limited cycle life. Therefore, a dead-band function can be added after the filter. Then, the ESS will not be in charging or discharging mode when the fluctuation is very small. The size of the dead-band should be optimized so that an optimal trade-off between output power quality and battery life can be achieved. Furthermore, the converter of the ESS has a power rating. So a saturation block is needed to limit the charging and discharging references to be in the safe range.

The control strategy is shown in Figure 2.11. The output P_{ESS} is the reference for the interfacing device of ESS. When $P_{ESS} > 0$, the ESS is discharging power to the microgrid. When $P_{ESS} < 0$, the ESS is absorbing

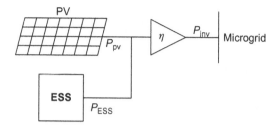

Figure 2.12 A PV microgrid for ramp-rate control.

the redundant energy from the PV system. If the dynamics of the ESS are fast, the output of the PV system can be smoothed satisfactorily.

2.5.3 Ramp-rate Control

In renewable microgrid, the output of the renewable sources (i.e., PV and wind turbine) may change rapidly. For example, the output of PV may have sudden drop when clouds come. Ramp-rate control is needed to satisfy the requirements for regulation and load following. The duty of ramp-rate control is to slow down the variation of renewable system's output and keep it at the desired rate of change (ramp-rate). In ramp-rate control, the ESS works as a power compensator in the microgrid.

To illustrate the ramp-rate control, an example PV microgrid with ESS [7] is presented in Figure 2.12. Suppose that the inverter in the microgrid has an efficiency η. So the output of the inverter to the microgrid can be calculated by (2.13).

$$P_{\text{inv}} = \eta \times (P_{\text{pv}} + P_{\text{ESS}}) \tag{2.13}$$

By taking derivative of the two sides of (2.13), the following equation can be obtained.

$$\frac{dP_{\text{inv}}}{dt} = \eta \times \left(\frac{dP_{\text{pv}}}{dt} + \frac{dP_{\text{ESS}}}{dt} \right) \tag{2.14}$$

This equation determines the ramp-rate of the inverter. Because the ESS is a controllable source, the control strategy for the desired output characteristic can be achieved by rewriting (2.14) as follows:

$$\frac{dP_{\text{ESS}}}{dt} = \frac{1}{\eta} RR_{\text{inv}} - \frac{dP_{\text{pv}}}{dt} \tag{2.15}$$

where RR_{inv} is the desired ramp-rate of the inverter output, that is, $RR_{\text{inv}} = (dP_{\text{inv}}/dt)|_{\text{des}}$. Note that the desired ramp-rate RR_{inv} is positive when the ramp-rate of PV is positive and is negative when PV ramps down. The characteristic of RR_{inv} can be described by

$$RR_{inv} = sgn\left(\frac{dP_{pv}}{dt}\right) \cdot |RR_{inv}| \qquad (2.16)$$

Therefore, by using Eq. (2.15), the required ramp-rate of ESS, dP_{ESS}/dt, is obtained. However, if the ramp-rate of renewable source is small, there is no need to regulate the output. Otherwise, the stability of the system may be affected. For example, if the output of PV is increasing with a ramp-rate (absolute value) lower than $\frac{1}{\eta}RR_{inv}$, the ESS will release energy to drive the ramp-rate of inverter to RR_{inv}, which makes the output ramp-rate even bigger, which is not desired. So a dead-band function should be employed.

$$D\left(\frac{dP_{pv}}{dt}\right) = \begin{cases} 0, & \text{if } |RR_{inv}| \geq \eta\left|\dfrac{dP_{pv}}{dt}\right| \\ 1, & \text{otherwise} \end{cases} \qquad (2.17)$$

Hence, with a dead-band in the controller, the desired ramp-rate RR_{inv} becomes the maximum allowable ramp-rate (MARR), because the ESS participates in the ramp-rate regulation only if the ramp-rate (absolute value) of the inverter is larger than $|RR_{inv}|$, thus limiting the ramp-rate under RR_{inv}.

However, the above dead-band function has a drawback. Assuming that the output of PV suddenly goes up to another value (like a step function), then the ramp-rate (derivative) of the PV output is like an impulse.

As Figure 2.13 shows, when the ramp-rate of PV goes beyond the dead-band at t_1, the ESS starts to absorb energy, so the ramp-rate of the inverter can be reduced. Until t_2, the output of PV reaches a new steady state, and the ramp-rate of PV falls below the dead-band then to zero. According to (2.17), the ESS will stop working and the PV will output directly to the inverter, which causes the output of inverter to suddenly step up. To solve this problem, the dead-band equation (2.17) can be modified as (2.18).

$$D'\left(\frac{dP_{pv}}{dt}\right) = \begin{cases} 0, & \text{if } |RR_{inv}| \geq \eta\left|\dfrac{dP_{pv}}{dt}\right| \text{ and } |P_{inv} - P_{pv}| \leq \varepsilon \\ 1, & \text{otherwise} \end{cases} \qquad (2.18)$$

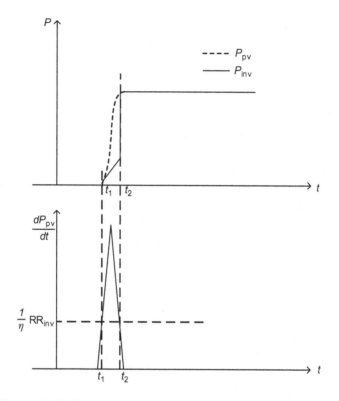

Figure 2.13 The naïve dead-band issue.

where ε is the tolerance value. For the modified dead-band (2.18), the ESS will stop working only if the ramp-rate of PV is small, and the output of inverter reaches PV's output. According to Euler's method, the output reference of ESS can be obtained by (2.19).

$$P_{\text{ESS}}(k) = \left[P_{\text{ESS}}(k - \Delta t) + \frac{dP_{\text{ESS}}}{dt} \Delta t \right] \cdot D' \left(\frac{dP_{\text{pv}}}{dt} \right) \qquad (2.19)$$

Figure 2.14 shows a ramp-rate control example. The output of PV starts to decline at time t_1 because the PV cells are shaded by clouds, and the ramp-rate control begins to command the ESS to support the ramp-rate of inverter. Until t_2, the output of PV goes up, so the ESS starts to charge and slow down the ramp up process. At t_3, the output of inverter matches the output of PV, so there is no need for ESS and thus can be turned off safely.

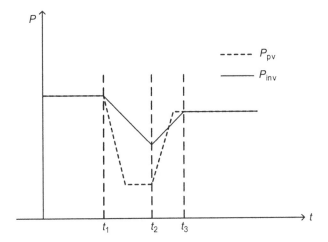

Figure 2.14 An example of ramp-rate control.

2.6 UNINTERRUPTIBLE POWER SYSTEM (UPS)

Uninterruptible Power Systems (UPS) are another important application of ESSs in renewable microgrids, especially for the islanded renewable microgrids. Renewable sources, such as solar panels and wind turbines, may suddenly stop generating power due to clouds and shading, nightfall or lack of wind. However, start-up of backup generators may take about 10 s [8]. During this period, outages may occur in the microgrid. However, if there is ESS in the microgrid, the loads can be supplied by ESS to ride through the power shortage.

Figure 2.15 shows an example of a PV microgrid. It has a diesel generator as the backup power and an ESS as the UPS. There is a voltage detector installed at the output of the PV. The voltage detector senses the output voltage of the PV. If the solar panels stop generating power, the output voltage will drop to zero. Then, the voltage detector will send a command to the backup generator and ESS immediately. The switches S_1 and S_2 will close, so the ESS can replace the PV as the power source to supply the microgrid, and in the meantime the diesel generator will start up. When the start-up process is completed, the diesel generator begins to supply the microgrid. Now, the ESS can be removed from the microgrid by breaking the switch S_1.

The PV may start to generate power again. So the diesel generator can be shut off when the output of PV is stable to supply the microgrid.

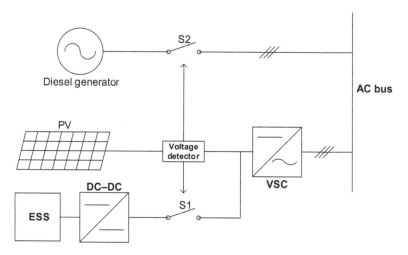

Figure 2.15 UPS for PV microgrid.

Then, the microgrid goes back to the normal operation. Since the ESS was discharged before, its SOC is not at the maximum. Thus, the microgrid controller can close switch S_1 and the ESS is charged by PV.

Note that if the outage lasts for a very short period, the diesel generator should not start up. Otherwise, the generator may be turned on and off too frequently, which is damaging. So, the microgrid control system should be smart enough to judge if power shortage will last for a long period or not. For a short period of outage, the ESS can support the microgrid on its own, but for long outage periods the backup generator should be turned on.

2.7 LOW-VOLTAGE RIDE THROUGH

When some incidents happen in the microgrid or main grid, the voltage may suddenly drop and cause some DGs to disconnect and stop producing any power. If a certain DG in the microgrid is tripped because of the voltage drop, this may cause other DGs to disconnect from the grid and cascading power outage may result. Therefore, DGs, like wind turbines, are required to remain connected to the grid for a period during the voltage sag. This requires that wind turbines should have the capability to operate under low voltage conditions which is referred to as low-voltage ride-through (LVRT). By cooperating with ESS, the LVRT ability of wind generators can be improved [9,10].

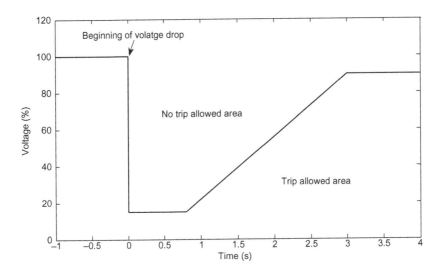

Figure 2.16 A typical LVRT requirement.

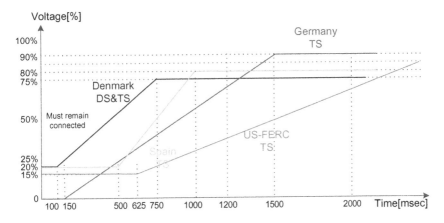

Figure 2.17 LVRT requirements for different countries [11].

The requirement of LVRT for a typical voltage sag event is shown in Figure 2.16. After the voltage drop happens, if the voltage is in the "no trip allowed area," the generator should not disconnect from the grid. If it is in the "trip allowed area," then the generator can be tripped. The requirements of LVRT are diverse in different countries [11]. Figure 2.17 shows the variety of LVRT requirements for different countries.

In the voltage sag event, the grid-side converter of the wind turbine cannot transport full power from the generator to the grid due to the voltage sag [12]. So ESS can be employed to absorb the redundant energy that the wind turbine generates. Thus, the control reference for the ESS can be,

$$P_{\text{ESSref}} = P_{\text{WT}} - P_{\text{grid}} \qquad (2.20)$$

where P_{WT} is the power generated by the wind turbine, P_{grid} is the allowable power injection from the wind turbine into the grid. Therefore, when the voltage drop has been detected, the ESS can be activated to perform LVRT feature, until the generator is allowed to trip. The structure of the wind turbine with ESS is the same as in Figure 2.8.

2.8 PLACEMENT OF THE ESS TO IMPROVE POWER QUALITY

The high penetration of renewable energy presents a challenge for the stable operation of microgrids. Besides the control methods discussed above to solve the power intermittency issue, the placement of ESS in a microgrid is also important for power quality. The voltage stability and continuity of service are two important aspects of power quality. So, the placement of ESS should take into account these factors. In this section the method to determine the voltage regulation ability and power quality indices will be introduced. And this information can be used to select the optimal placement of ESS [13,14].

First, the method to realize the voltage regulation ability of ESS can be derived from the power flow problem. In a power system, the power balance equation for real and reactive power of bus i can be computed by:

$$P_i = |V_i| \sum_{j=1}^{N} |V_j| Y_{ij} \cdot \cos(\theta_{ij} - \delta_i + \delta_j) \qquad (2.21)$$

$$Q_i = |V_i| \sum_{j=1}^{N} |V_j| Y_{ij} \cdot \sin(\theta_{ij} - \delta_i + \delta_j) \qquad (2.22)$$

where P_i and Q_i are the real power and reactive power injection at bus i. V_i is the voltage amplitude at bus. Y_{ij} is the elements of bus admittance matrix. θ_{ij} denotes the angle of admittance between bus i and bus j. δ_i and δ_j are the voltage angles at bus i and bus j, respectively. By taking derivative of the above equations with respect to voltage angles and voltage magnitudes, the Newton–Raphson power flow equation can be obtained,

$$\begin{bmatrix} \Delta P \\ \Delta Q \end{bmatrix} = J \begin{bmatrix} \Delta \delta \\ \Delta V \end{bmatrix} \tag{2.23}$$

where J is the Jacobian matrix. It has the following form:

$$J = \begin{bmatrix} \dfrac{\partial P}{\partial \delta} & \dfrac{\partial P}{\partial V} \\ \dfrac{\partial Q}{\partial \delta} & \dfrac{\partial Q}{\partial V} \end{bmatrix} \tag{2.24}$$

Equation (2.23) describes the power mismatch when the voltage is changed. In order to determine the voltage sensitivity with respect to the power change, Eq. (2.23) can be rewritten, and the following relationship is achieved.

$$\begin{bmatrix} \Delta \delta \\ \Delta V \end{bmatrix} = J_{\text{inv}} \begin{bmatrix} \Delta P \\ \Delta Q \end{bmatrix} \tag{2.25}$$

where the matrix J_{inv} is "inverse" Jacobian matrix as defined below,

$$J_{\text{inv}} = \begin{bmatrix} \dfrac{\partial \delta}{\partial P} & \dfrac{\partial \delta}{\partial Q} \\ \dfrac{\partial V}{\partial P} & \dfrac{\partial V}{\partial Q} \end{bmatrix} = \begin{bmatrix} \dfrac{\partial \delta_2}{\partial P_2} & \cdots & \dfrac{\partial \delta_2}{\partial P_n} & \dfrac{\partial \delta_2}{\partial Q_2} & \cdots & \dfrac{\partial \delta_2}{\partial Q_n} \\ \vdots & \ddots & \vdots & \vdots & \ddots & \vdots \\ \dfrac{\partial \delta_n}{\partial P_2} & \cdots & \dfrac{\partial \delta_n}{\partial P_n} & \dfrac{\partial \delta_n}{\partial Q_2} & \cdots & \dfrac{\partial \delta_n}{\partial Q_n} \\ \dfrac{\partial |V_2|}{\partial P_2} & \cdots & \dfrac{\partial |V_2|}{\partial P_n} & \dfrac{\partial |V_2|}{\partial Q_2} & \cdots & \dfrac{\partial |V_2|}{\partial Q_n} \\ \vdots & \ddots & \vdots & \vdots & \ddots & \vdots \\ \dfrac{\partial |V_n|}{\partial P_2} & \cdots & \dfrac{\partial |V_n|}{\partial P_n} & \dfrac{\partial |V_n|}{\partial Q_2} & \cdots & \dfrac{\partial |V_n|}{\partial Q_n} \end{bmatrix}$$

$$\tag{2.26}$$

Eq. (2.25) represents the relationship between the power variation and voltage change in the power system. Hence it is easy to calculate the voltage magnitude and angle deviation when the real power and reactive power changes. From this equation, the voltage regulation ability of ESS can be understood. Also, the optimal location to install the ESS to have the biggest voltage regulation ability can be found with this equation. At this optimal bus, the ESS can use the lowest power to achieve the voltage regulation goal.

Besides the voltage regulation ability, several power quality indices addressing the continuity of service should be considered when deciding the location of ESS. They are System Average Interruption Duration Index (SAIDI), System Average Interruption Frequency Index (SAIFI) and Customer Average Interruption Duration Index (CAIDI). These indices are widely used to evaluate power quality in power systems. SAIDI describes the average outage duration for each customer served. It can be calculated through Eq. (2.27).

$$SAIDI = \frac{T_{int}}{N_{cs}} \qquad (2.27)$$

where T_{int} is total duration of customer interruptions, and N_{cs} is the total number of customers served. SAIDI is usually computed on the basis of a whole year's data, measured at minute level.

SAIFI is the average number of interruptions that a customer would experience. It can be described as follows:

$$SAIFI = \frac{N_{int}}{N_{cs}} \qquad (2.28)$$

where N_{int} is the total number of customer interruptions. SAIFI is also computed on the basis of one year's data.

CAIDI is calculated by:

$$CAIDI = \frac{SAIDI}{SAIFI} \qquad (2.29)$$

The above indices provide useful information to construct microgrids containing an ESS. By simulating different microgrid configuration with the whole year's data, the optimal configuration of microgrid with ESS to gain the best power quality can be obtained.

2.9 VOLTAGE REGULATION USING ESS

Here, PQ decoupling control is adopted for the ESS. Hence, it is able to support microgrid voltage through reactive power control. Note that numerous voltage support equipment and means are available including

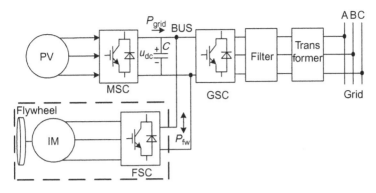

Figure 2.18 PV-Flywheel inverter-based distributed generation system [16].

excitation system, tap changer in transformer, capacitor, SVC and STATCOM. Furthermore, the voltage support by ESS will affect the capacity of its active power regulation capacity. Therefore, voltage support by ESS should be used only as an auxiliary function.

Voltage regulation using energy storage has been researched in several studies. In Ref. [15], in order to improve the dynamic and transient voltage stability of a microgrid including a synchronous generator and an asynchronous wind turbine, the voltage stability control strategy based on the small-signal model of the supercapacitor energy storage (SES) is presented. A control method for voltage regulation using flywheel energy storage is introduced in Ref. [16]. Flywheels are employed as energy storage medium, exhibiting high performance with electrical applications such as power conditioning, frequency regulation and voltage sag compensation due to their capability of storing energy in the form of kinetic energy, which is a function of the rotating speed and mass [17,18].

The block diagram for the FESS along with the PV-inverter-based distributed generation is shown in Figure 2.18 [16]. The flywheel is coupled to a low speed squirrel cage induction machine (IM), which is connected to a DC bus through a FESS converter. The FESS is considered very suitable for power conditioning applications due to its robustness during severe variations in the operating mode [19–22]. The d-q model of the induction machine [22,23] and the FESS side converter is presented in Eqs (2.30)–(2.35).

$$V_{ds} = r_s i_{ds} + p\lambda_{ds} - \omega_e \lambda_{qs} \qquad (2.30)$$

$$V_{qs} = r_s i_{qs} + p\lambda_{qs} - \omega_e \lambda_{ds} \qquad (2.31)$$

$$P_s = \frac{3}{2}(v_{ds}i_{ds} + v_{qs}i_{qs}) = P_{CU_{stator}} + P_{CU_{rotor}} + T_m\omega_m \qquad (2.32)$$

$$Q_s = \frac{3}{2}(v_{ds}i_{qs} - v_{qs}i_{ds}) = L_m\omega_e i_{ds}^2 \qquad (2.33)$$

$$T_m = \frac{3}{2}\frac{PL^2}{L_r}i_{qs}i_{ds} \qquad (2.34)$$

$$P_{fw} = P_s \qquad (2.35)$$

The amount of available power in the flywheel is given by (2.36).

$$P_{fw} = P_{pv} - P_{grid} \qquad (2.36)$$

The amount of stored energy in the flywheel can be expressed using (2.37).

$$E = \frac{1}{2}J\left(\omega_{cn}^2 - \omega_{disch}^2\right) = P_{fw}t_{disch} \qquad (2.37)$$

where J is the flywheel moment of inertia, ω_{ch} and ω_{disch} are the charging and discharging speeds respectively and t_{disch} is the discharging time. The desired quantity to be controlled is the torque component of current (quadrature current i_{qs}), whose reference value is generated depending on the actual terminal voltage of the renewable DG (PV in this discussion) in case it exceeds the threshold value for safe operation. The reactive power component is responsible for the generation of the direct-axis current component.

Voltage support by ESS should be activated only when voltage level is in an emergency situation. Taking advantage of its rapid speed of output regulation, the reactive power can be output by ESS firstly to support the voltage. As the voltage returns to the normal range, reactive power output should be reduced even though the voltage is still below the ideal value so that more active power regulation capacity is retained in the ESS.

The active power regulation capacity of ESS is affected directly by the reactive power output. Active power regulation requirement for ESS can be predicted using wind power prediction technology. Hence, the upper limit can be calculated so that the required active power regulation capacity for ESS can be guaranteed. The voltage regulation strategy for ESS is shown in Figure 2.19. Voltage regulation is activated only when the voltage at the ESS connection is beyond the

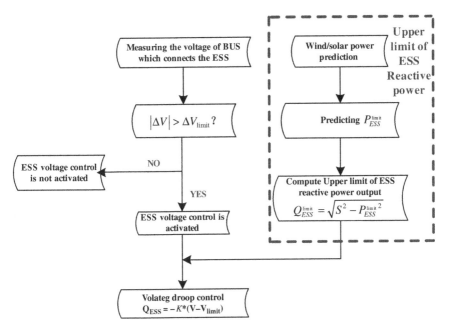

Figure 2.19 The voltage regulation strategy for ESS participation.

threshold value, which means the voltage is in the critical range. The upper limit of ESS reactive power output is decided based on the most severe wind power fluctuation situation so that the required active power regulation capacity is guaranteed. The details of each part in Figure 2.19 are introduced in the following sections.

2.9.1 The Threshold Value for ESS Voltage Regulation Activation

For microgrid operation, all the bus voltages are required to be within the normal range. The voltage deviation threshold value ΔV_{limit} is set to decide whether the voltage is near the critical level and whether the voltage control by ESS should be activated. When the voltage deviation is less than ΔV_{limit}, voltage is considered to be in the normal range. Only normal regulation, such as exciters, transformers tap changers, capacitors, SVC and STATCOM, should be used for the voltage control while the reactive power regulation capacity of ESS is reserved. When the voltage deviation is bigger than ΔV_{limit}, voltage is in the critical level or emergency condition and the reactive power regulation by ESS is activated then.

ΔV_{limit} is also decided based on the power quality required by the loads. For critical load or load which are sensitive to the voltage, ΔV_{limit} should be set smaller, for example, 0.05 p.u. While for common load, ΔV_{limit} can be set higher, for example, 0.1 p.u.

2.9.2 Upper Limit of ESS Reactive Power Output
The apparent power S for the installed ESS is a fixed value depending on its overall rating. Hence, the active power capacity of ESS will be affected directly by its reactive power regulation output. When the upper limit of ESS reactive power output is set based on active power requirement for the ESS, the active power regulation capacity can be guaranteed. The relationship between the maximum active power output $P_{\text{ESS}}^{\text{limit}}$ and the maximum reactive power output $Q_{\text{ESS}}^{\text{limit}}$ is expressed by:

$$\left(P_{\text{ESS}}^{\text{limit}}\right)^2 + \left(Q_{\text{ESS}}^{\text{limit}}\right)^2 = S^2 \tag{2.38}$$

The frequency regulation capacity expected from ESS depends on wind power fluctuation. Wind power prediction technology is able to provide the power disturbance due to wind power fluctuation so that the active power regulation capacity requirement for ESS can be calculated. For ESS, active power regulation capacity should be guaranteed as the first priority while the reactive power support is the auxiliary function. Hence, the reactive power output upper limit of ESS can be optimized dynamically using

$$Q_{\text{ESS}}^{\text{limit}} = \sqrt{S^2 - \left(P_{\text{ESS}}^{\text{limit}}\right)^2} \tag{2.39}$$

where $P_{\text{ESS}}^{\text{limit}}$ is revised in real time based on the wind power prediction information. The required active power regulation capacity $P_{\text{ESS}}^{\text{limit}}$ can be guaranteed when the reactive power output upper limit of ESS $Q_{\text{ESS}}^{\text{limit}}$ is set by (2.39). Besides, $P_{\text{ESS}}^{\text{limit}}$ should be decided based on the most severe wind power fluctuation situation so that there is enough allowance of $P_{\text{ESS}}^{\text{limit}}$.

2.9.3 Reactive Power Support by ESS
When the voltage level is in the emergency range, the auxiliary reactive power control by ESS should be activated to support the voltage. Reactive power droop control strategy by ESS is adopted here so that reactive power can be injected to the microgrid based on

the reactive power requirement. The ESS reactive power output Q_{ESS} can be calculated by

$$Q_{\text{ESS}} = -K \cdot (V - V_{\text{limit}}) \qquad (2.40)$$

where Q_{ESS} is ESS reactive power output; K is the reactive power–voltage droop coefficient; V is the voltage of the bus which is connected to the ESS; V_{limit} is the threshold value for ESS voltage regulation activation, which can be expressed by

$$V_{\text{limit}} = \begin{cases} 1 + \Delta V_{\text{limit}}, & V > 1 \\ 1 - \Delta V_{\text{limit}}, & V < 1 \end{cases} \qquad (2.41)$$

Based on (2.40) and (2.40), ESS absorbs the reactive power when voltage is higher than the upper limit of the threshold value so that bus voltage can decrease. On the other hand, ESS provides the reactive power support when voltage is lower than the upper limit of the threshold value so that bus voltage can increase.

In addition, the reactive power support is provided based on the voltage deviation magnitude when (2.40) is adopted. More reactive power support is provided when the voltage deviation becomes larger. Hence, the reactive power can be injected into the microgrid in proportion to the voltage deviation based on this control strategy.

2.9.4 Automatic Substitution of ESS Reactive Power Regulation Capacity

When voltage level returns to normal level, it indicates that the regular reactive regulation device is able to support the bus voltage. Based on the presented reactive power control strategy of ESS, the reactive power output of ESS should be replaced so that more active regulation capacity of ESS becomes available. Note that even if the voltage is still higher or lower than the ideal value within normal range, since the active power regulation is the major function of ESS, its reactive power output should be released when the voltage returns to the normal level.

This control target can be realized based on (2.41). Reactive power output of ESS becomes smaller as the voltage returns to the normal range. The ESS reactive power output is equal to zero and thus the ESS reactive power regulation function is de-activated when the voltage is equal to the threshold value V_{limit}. In this way, the switching between activation and de-activation can be done smoothly.

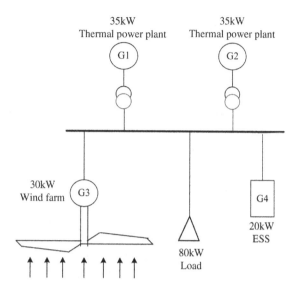

Figure 2.20 The structure of the simulation system.

2.9.5 The Reactive Power Control Strategy Verification Case

This section presents Real Time Digital Simulator (RTDS) simulation results to demonstrate the effectiveness of the proposed control scheme. The installed capacity of the test system is 120 KW, including two 35 KW thermal power units, one 30 KW wind turbine generator and one 20 KW ESS. The total load of this system is 80 KW and 36 KVar. The structure of the system is shown in Figure 2.20. Note that the test system nominal frequency is 50 Hz.

In this simulation case, 19 KVar reactive power disturbance is set at $t = 2s$. A comparison study is done to demonstrate the control effect of the presented ESS reactive power control strategy on the voltage sag. Note that the proposed voltage regulation by ESS is done if active power regulation capacity of ESS is guaranteed.

The comparison of voltage response when reactive power disturbance happens is shown in Figure 2.21. As shown in the solid curve, the bus voltage V_{BUS} falls to 0.833 p.u. and stabilizes at 0.928 p.u. without the voltage regulation of ESS participation. Since the voltage regulation of ESS is not activated, reactive power output of ESS is equal to zero when the voltage drops, as shown in the solid curve in Figure 2.22.

The voltage response with the voltage regulation of ESS participation is shown as the dotted curve in Figure 2.21. ΔV_{limit} is set as 0.05 p.u. in

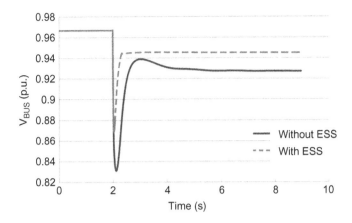

Figure 2.21 The voltage response when reactive power disturbance happens (1) without the voltage regulation of ESS participation (2) with the voltage regulation of ESS participation.

this simulation case. Based on the reactive power control strategy of ESS shown in Figure 2.19, voltage regulation of ESS is activated to support the bus voltage according to (2.40) when the voltage is lower than 0.95 p.u. Based on (2.39), the upper limit of the reactive power output is calculated as 12 KVar when the maximum active power output is equal to 16 KW based on the wind power prediction.

The reactive power output of ESS is shown as the dotted curve in Figure 2.22. Reactive power output of ESS increases rapidly when the bus voltage falls. Its maximum value is 11.5 KVar in the transient process and its stable value is 8.2 KVar. Owing to the voltage regulation of ESS, the voltage is 0.943 p.u. in the steady state, which is 0.015 p.u. higher than the one without the ESS participation. The maximum voltage deviation also decreases by 0.038 p.u. thanks to the voltage regulation of ESS. Note that the upper limit of ESS reactive power output is updated in real time so that the reactive power regulation of ESS will not affect its active power regulation capacity. That is to say, the reactive power regulation is realized while the investment cost of ESS will not increase for this auxiliary function.

2.10 ESS AS SPINNING RESERVE

The spinning reserve is the amount of unused capacity in online energy assets which can compensate for power shortages or frequency drops within a given period of time. Traditionally, the spinning reserve is a

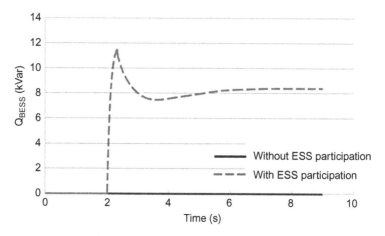

Figure 2.22 The reactive power output of ESS when reactive power disturbance happens (1) without the voltage regulation of ESS participation (2) with the voltage regulation of ESS participation.

concept for large synchronous generators. If the largest generator in the power system is tripped, the remaining generators should increase their output to recover the power shortage. So, some generators in the power system are required to operate below their rated value to prepare for unexpected demand or contingencies. The more reserve is needed, the operating points of generators are further away from the rated value and thus are less efficient. Also, if the generators work under inefficient conditions, CO_2 and other emissions will increase.

Nowadays, due to the high penetration of renewable energy in microgrids, its power reliability is lower than the traditional power system. Energy generation is heavily reliant on weather conditions, which change constantly. As a result, power shortages may happen more often in microgrids. So, the issue of spinning reserve is more important for microgrid operation.

Luckily, the ESS, as the commonly used device in microgrid, can be used to accommodate some of the spinning reserve requirement for microgrid generators. With the help of an ESS, the generators can work more closely to their rated values. Also, the ESS can react faster than many generators, so the power shortage can be recovered in a quicker fashion.

In spinning reserve technology, the most important problem is how much capacity should be reserved. Both the reliability and the economics of the microgrid should be considered when deciding on the amount of spinning reserve. It is necessary to set the reserve capacity at optimal value, so the operation of microgrid is most economical. Hence, this problem is usually handled by an optimization method [24]. An example objective function of the optimization problem is stated in (2.42).

$$\max(RN - OC - RC) \tag{2.42}$$

The objective function describes the profit of the microgrid, which the operators desire to maximize. In function (2.42), RN is the total revenue of the microgrid as power from DGs are sold to users in the microgrid. OC indicates the operational cost, which includes the running cost and start-up cost of each DG and RC is the total cost for spinning reserve. By dividing the operating process of microgrid into several time periods (e.g., divide one day into T periods). These three terms can be formulated as follows:

$$RN = \sum_{t=1}^{T} \sum_{i=1}^{I} PR(t) P_i^t \tag{2.43}$$

where t is the current time period among all the T periods. I denotes the total number of DG units within the microgrid. $PR(t)$ is the market price of active power during period t; P_i^t is output power of unit i at period t.

$$OC = \sum_{t=1}^{T} \sum_{i=1}^{I} \left[C_i^t \left(P_i^t, S_i^t \right) + SC_i \cdot K_i^t \right] \tag{2.44}$$

where $C_i^t(*)$ is the running cost function of DGi, which is a function of output power and status of unit i, S_i^t. S_i^t can be either 1 or 0: If the unit i is online, then $S_i^t = 1$, otherwise $S_i^t = 0$. SC_i is the start-up cost of unit i. And K_i^t is the start-up status of unit i; it is 1 if the DG starts up during period t and 0 otherwise.

$$RC = \sum_{t=1}^{T} \sum_{i=1}^{I} q_i^t R_i^t + EENS \cdot VOLL \tag{2.45}$$

where q_i^t is reserve price of unit i during period t; R_i^t is reserve power of unit i during period t. Expected Energy Not Supplied (EENS) represents the Expected Energy Not Supplied, and VOLL is the Value of Lost Load. VOLL describes the average expense of consumers for the accidental loss of 1 kWh of electricity. These data usually come from the consumer surveys [25]. The EENS can be calculated in different ways, considering uncertainty of wind power, solar power and the load.

In addition to the objective function, the optimization problem needs to satisfy some constraints. The power balance constraint is the essential requirement for a power system, as given in Eq. (2.46). Since the optimization is conducted before the actual operation and the load of microgrid is unknown, the forecasted load P_D^t can be used.

$$\sum_{i=1}^{I} P_i^t = P_D^t \quad \forall t = 1, \ldots, T \tag{2.46}$$

Next, the Loss of Load Probability (LOLP) constraint requires that the probability should be lower than a maximum value in every time period.

$$\text{LOLP}^t \leq \text{LOLP}_{\text{max}}^t \tag{2.47}$$

where LOLP^t is the LOLP of period t, and the $\text{LOLP}_{\text{max}}^t$ indicates its maximum value.

Also, the following spinning reserve constraint should be satisfied.

$$\begin{cases} R_i^t \leq P_i^{\text{max}} S_i^t - P_i^t \\ R_i^t \leq S_i^t \cdot \text{RR}_i^{\text{up}} \cdot \tau \end{cases} \tag{2.48}$$

In the inequalities, P_i^{max} is the maximum output of unit i. RR_i^{up} is the maximum ramp up rate of unit i. τ is the available time interval for the generators to ramp up their output to deliver the reserve energy. The first inequality in (2.48) says the sum of output power P_i^t and reserve R_i^t should be lower than the maximum output ability of unit i, which is $P_i^{\text{max}} S_i^t$. The second term represents that the reserve should meet the ramp-rate requirement.

The above constraints are the general requirements for all generation units. But for the specific ESS unit, additional requirements are necessary:

$$P_{\text{ESS}}^t \leq P_{\text{ESS,max}} \tag{2.49}$$

$$C_{\text{min}} \leq C^t \leq C_{\text{max}} \tag{2.50}$$

The first constraint limits the output power of ESS P_{ESS}^t at period t to be below the maximum value $P_{\text{ESS,max}}$. The second inequality indicates the maximum and minimum value of the stored energy C^t. The stored energy of ESS can be calculated by the charging and discharging equation (2.51).

$$\begin{cases} C^{t+1} = C^t + dt \cdot \dfrac{P_{\text{ESS}}^t}{\eta}, \quad P_{\text{ESS}}^t < 0 \\ C^{t+1} = C^t + dt \cdot P_{\text{ESS}}^t \cdot \eta, \quad P_{\text{ESS}}^t > 0 \end{cases} \tag{2.51}$$

where dt is the duration of period t (e.g., 15 min), and η is the efficiency of ESS and its inverter. In (2.51), when P^t_{ESS} is negative, it means the ESS is delivering power to the microgrid. On the other hand, ESS is being charged if P^t_{ESS} is positive.

By solving the optimization problem with above constraints, the optimal spinning reserve value for the microgrid to achieve the maximum profit can be obtained.

2.11 CASE STUDY: OPERATING RESERVES USING ESS

This case study focuses on utilizing ESSs to provide operating reserves (spinning and non-spinning reserves). This is possible because the ESS can quickly and accurately perform real time balancing of supply and demand.

2.11.1 Problem Formulation

The operation of the microgrid is modeled via a security constrained unit commitment (SCUC) and security constrained economic dispatch (SCED). The real time balancing of load and generation is performed via automatic generation control (AGC). This portion of the controller is tasked with setting generator operating set-points every 4 s in order to correct the area control error (ACE) realized in the system. Here, the ACE is a measure of the current imbalance between generation and load occurring in the microgrid every 4 s taken mathematically as:

$$ACE(t) = \sum_I P_{i,t-1} - \sum_K \sum_B D_{k,b,t-1} \qquad (2.52)$$

In (2.52), the ACE during the current time interval is based on the imbalance occurring in the previous time interval, in this case, the interval 4 s previous to the current interval. The sets of I, K, and B are the sets of generation resources, load points, and buses throughout the microgrid respectively. Once the ACE is calculated, updated generator set-points based on the regulation reserve schedules as determined by the economic dispatch solution are calculated and sent to all generation resources. By co-optimizing generation and ancillary services, the most efficient set of generator set-points is calculated and dispatched.

In addition to the traditional unit commitment formulation that contains generator output limits, minimum run/down times, generator ramp rates, load balance, transmission constraints, and ancillary service constraints, similar to Ref. [26], ESS constraints that capture their operating characteristics are included. The following equations detail the modeling of ESS.

$$\sigma_{i,t} = \sigma_{i,t-1} - [P_{i,t} \cdot \eta_{\text{dis},i} + PS_{i,t} \cdot \eta_{\text{chg},i}] \tag{2.53}$$

where σ represents the storage level, i represents the index of the ESS unit, t represent the current time index, $P_{i,t}$ is the current generation schedule of unit i at time t, $PS_{i,t}$ is the current charging schedule of unit i at time t, and $\eta_{\text{dis},i}$ and $\eta_{\text{chg},i}$ are the discharging and charging efficiency of the ESS unit respectively. This equation is used to model the current level of the ESS based on the previous level and current action of the ESS unit. Note that either $P_{i,t}$ or $PS_{i,t}$ or both must be zero at all times because the battery can be in either discharging, charging or idle mode.

If a final storage level is required, the following constraint is considered:

$$\sigma_{i,\text{end}} = \alpha + \rho^{+} - \rho^{-} \tag{2.54}$$

where $\sigma_{i,\text{end}}$ is the storage level of unit i at the end of the optimization horizon, α is a previously specified final storage level, and ρ^{+} and ρ^{-} are positive and negative deviations from that final storage level respectively. The positive and negative deviations are associated with respective penalty factors to ensure that the optimization tries to schedule the ESS to meet the required final storage level.

The following constraint is used to ensure that the storage level does not exceed its maximum capacity at the end of each optimization.

$$\sigma_{i,t} \leq \sigma_{\text{max},i} + \sigma_{\text{wasted}} \tag{2.55}$$

where $\sigma_{i,t}$ is the storage level of unit i at time t, $\sigma_{\text{max},i}$ is the maximum storage level of unit i, and σ_{wasted} is any wasted storage capacity (i.e., energy that is trying to be stored beyond the capacity limit that is ultimately wasted).

Equation (2.56) is the operating mode exclusivity constraint:

$$I_{i,t} + IP_{i,t} \leq 1 \tag{2.56}$$

where $I_{i,t}$ is the unit generating commitment status at time t, and $IP_{i,t}$ is the charging commitment status at time t. These binary variables are used to determine which mode the ESS units are operating in. This constraint ensures that a unit cannot be both generating and storing energy at the same time.

The following constraint ensures that the unit does not exceed its storage capacity limits during real time operation:

$$PS_{\min,i} \cdot IP_{i,t} \le PS_{i,t} \le PS_{\max,i} \cdot IP_{i,t} \qquad (2.57)$$

where $PS_{\min,i}$ is the minimum required charging output of unit i, $PS_{\max,i}$ is the maximum allowable charging output of unit i, $IP_{i,t}$ is the charging commitment status of unit i at time t, and $PS_{i,t}$ is the charging schedule of unit i at time t.

The following two constraints ensure that the ESS units do not exceed their capacity while providing spinning reserves in charging mode.

$$PS_{i,t} - \sum_{\substack{\Gamma = \text{Up} \\ \text{Reserves}}} R_{i,t,\Gamma} \ge IP_{i,t} \cdot PS_{\min,i} - I_{i,t} \cdot P_{\max,i} \qquad (2.58)$$

$$PS_{i,t} + \sum_{\substack{\Gamma = \text{Down} \\ \text{Reserves}}} R_{i,t,\Gamma} \le IP_{i,t} \cdot PS_{\max,i} + I_{i,t} \cdot P_{\max,i} \qquad (2.59)$$

where $PS_{i,t}$ is the storage schedule of unit i at time t, $R_{i,t}$, is the reserve schedule of unit i at time t, Γ is the type of reserve product being scheduled (e.g., contingency spin, contingency non-spin, up regulation, down regulation, etc.), IP is the charging commitment status, I is the discharging mode commitment variable, $PS_{\min,i}$ is the minimum required storage level of unit i, and $P_{\max,i}$ is the maximum possible generating level of unit i. The second term on the right hand side of (2.58) and (2.59) are required to relax the constraints if the unit is not operating in charging mode.

The following constraint is used to enforce the minimum charging time of the ESS units.

$$\sum_{t}^{t+TP_{\text{on},i}-1} IP_{i,t} \ge TP_{\text{on},i} \cdot yp_{i,t} \cdot (1 - FO_{i,t}) \qquad (2.60)$$

where, $TP_{\text{on},i}$ is the minimum charging time requirement of unit i, $yp_{i,t}$ is the charging mode starting indicator which is 1 if the unit is starting to charge and zero otherwise, and $FO_{i,t}$ is an outage binary variable where a value of 1 represents unit i being taken offline at time t. Equation (2.60) essentially keeps track of consecutive charging commitment variables and ensures that the number of consecutive charging commitment variables is equal to the minimum charging time requirement of the ESS once a unit has started charging.

The following two constraints enforce the unit ramp-rate in the up and down directions respectively.

$$PS_{i,t} - PS_{i,t-1} \leq PRR_i \cdot (1 - yp_{i,t}) + PS_{\min,i} \cdot yp_{i,t} \qquad (2.61)$$

$$PS_{i,t} - PS_{i,t-1} \geq - PRR_i \cdot (1 - zp_{i,t}) + PS_{\min,i} \cdot zp_{i,t} \qquad (2.62)$$

where $PS_{i,t}$ is the charging schedule of unit i at time t, PRR_i is the charging ramp-rate of unit i, $yp_{i,t}$ is a charging start-up indicator, $zp_{i,t}$ is the shutting down equivalent of yp and indicates when the ESS is exiting charging mode, and $PS_{\min,i}$ is the minimum required charging level of unit i.

The following two constraints are used to model the binary start up and shut-down indicator variables:

$$yp_{i,t} - zp_{i,t} = IP_{i,t} - IP_{i,t-1} \qquad (2.63)$$

$$yp_{i,t} + zp_{i,t} \leq 1 \qquad (2.64)$$

Equation (2.63) is used to model the behavior of the start-up and shut-down ESS indicators. Equation (2.64) is used to ensure that a unit is not both charging and discharging at the same time.

The following constraint is used to model the start-up trajectory of units starting to store energy modeled after [27].

$$PS_{i,t} - \sum_{\substack{\Gamma = \\ \text{Up Reserves}}} R_{i,t,\Gamma} \geq \alpha + \beta - \gamma \qquad (2.65)$$

where

$$\alpha = PS_{\min,i} \cdot \left(IP_{i,t} - \sum_{H=t}^{t+\text{PDP}-1} zp_{i,H} - \sum_{H=t-\text{PUP}+1}^{t} yp_{i,H} \right) \qquad (2.66)$$

$$\beta = PS_{\min,i} \left(\sum_{H=t-\text{PUP}+1}^{t} (H - t + 1) yp_{i,t} \right) \min\left(1, \frac{1}{tp_{\text{start}}} \right) \qquad (2.67)$$

$$\gamma = P_{\max,i} \cdot I_{i,t} \qquad (2.68)$$

where $PS_{i,t}$ is the charging schedule of unit i at time t, PUP is the number of intervals a unit has been storing energy during start-up, PDP is the number of intervals a unit has been storing energy during

Table 2.1 System Characteristics	
Number of Thermal Generators	10
Nameplate Thermal Capacity	2380 kW
Total Renewable Energy Penetration	9.2%
Number of Network Buses	24
Number of Network Branches	40
Size of ESS	300 kWh

shut-down, $R_{i,t}$ is the reserve schedule of unit i at time t, yp is the charging startup indicator, tp_{start} is the time it takes for a unit to reach charging status, $P_{max,i}$ is the maximum discharge level of the unit, and $I_{i,t}$ is the generation commitment variable. Equation (2.66) is required to ensure that a unit is storing above the minimum required storage level if a unit is neither starting up nor shutting down. Equation (2.67) is used to set the trajectory limit of a unit during startup. The trajectory is assumed to be a linear trajectory over the amount of time it takes a unit to reach its minimum storage level. Equation (2.68) is required to relax the constraint if the unit is generating rather than storing energy.

2.11.2 Simulation Setup

The system simulated is a modified version of the single-area IEEE Reliability Test System (RTS96) to reflect a microgrid that might serve large industrial/commercial customers or even small towns. The modifications include building redundancy in the system in order to establish multiple paths of power to supply all loads, inspired by the perfect power microgrid at the Illinois Institute of Technology [28]. Distributed generators are connected throughout the microgrid. In additional to the distributed generators, distributed wind and solar generators are added to the system to mimic a typical microgrid configuration. The load, wind and solar data are based on available data from the National Renewable Energy Laboratory's Western Wind and Solar Integration Study Phase 2 dataset 0. Table 2.1 summarizes the system characteristics.

A one-line diagram of the system is shown in Figure 2.23.

The location of the wind and solar generators are selected to maximize access to transmission. The system is simulated for 4 weeks (the third week in January, April, July and October) in order to capture the

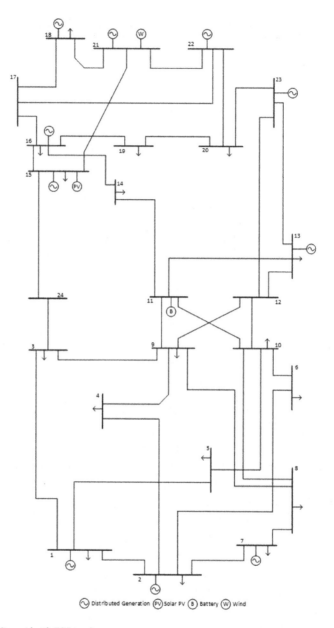

Figure 2.23 Microgrid with ESS topology.

seasonal behavior of wind and solar. Three cases are simulated to capture the benefit of energy storage. Case 1 does not include any ESSs. Case 2 includes ESSs participating in both energy and ancillary service provisions. Case 3 only allows the ESS to provide ancillary services.

Table 2.2 Simulation Results		Case 1	Case 2	Case 3
January	Cost ($M)	1.638	1.610	1.608
	AACEE (kWh)	597	632	479
	Sigma ACE (kW)	5.01	5.46	4.11
	MAACE (kW)	3.55	3.76	2.85
	VG Curtailment (kW)	314	169	243
April	Cost ($M)	1.336	1.303	1.289
	AACEE (kWh)	711	772	588
	Sigma ACE (kW)	6.35	6.72	5.37
	MAACE (kW)	4.24	4.59	3.50
	VG Curtailment (kW)	875	300	395
July	Cost ($M)	1.987	1.961	1.934
	AACEE (kWh)	422	488	368
	Sigma ACE (kW)	3.60	4.31	3.04
	MAACE (kW)	2.52	2.90	2.19
	VG Curtailment (kW)	435	73	221
October	Cost ($M)	1.451	1.485	1.468
	AACEE (kWh)	713	711	502
	Sigma ACE (kW)	6.13	6.35	4.31
	MAACE (kW)	4.24	4.23	2.99
	VG Curtailment (kW)	225	666	613

2.11.3 Simulation Results and Discussions

Table 2.2 summarizes the numerical results from the simulations performed in this study. As expected, in general, the addition of the ESS reduces the total system production cost. This is the cost for producing energy from all generation assets in the microgrid. In October, under low loading and high wind and solar output, the trends are reversed. The production is increased by including an ESS. Thermal generators are operating at or near their respective minimum generating levels. This results in the excess curtailment of variable generation (VG) wind and solar power along with the rise in production cost.

The AACEE is the absolute ACE in energy. This metric is a summation of the absolute value of ACE accrued throughout the study period and provides insight into how well the system is being balanced. Notice that including the ESS generally increases the absolute amount

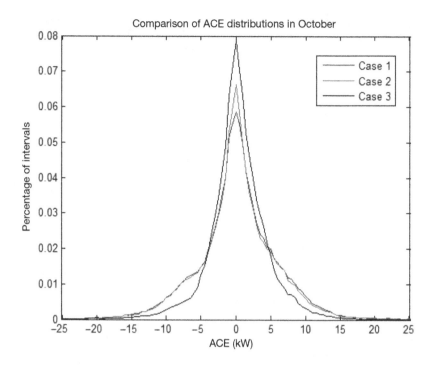

Figure 2.24 Distribution of ACE in October across all cases.

of ACE occurring in the system. This is most likely due to reducing the head room of online thermal generators and reducing their shutdowns. However, by allowing the ESS to focus only on providing ancillary services, particularly regulation reserves, the AACEE is reduced, even below the levels found in Case 1. This is because the ESS offers superior tracking abilities to the microgrid controller and is able to correct the ACE at a much faster pace than the thermal generators. A similar conclusion can be made for the sigma ACE and mean absolute area control error (MAACE) metrics. The sigma ACE is the standard deviation of the imbalance and provides insight into the variability of the ACE. The MAACE provides some insight in the average magnitude of the imbalance every 4 s.

Figure 2.24 shows the distribution of ACE across all three cases in October. Notice that the distribution becomes closer to 0 kW as the ESS is included in the microgrid and even tighter if the ESS focuses on providing only reserves. A similar behavior can be observed in the other weeks.

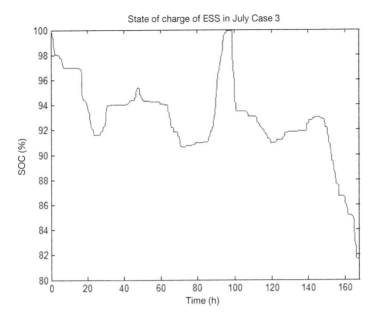

Figure 2.25 State of Charge of the ESS in July for Case 3.

Figure 2.25 shows the state of charge of the ESS in July for Case 3. By allowing the ESS to focus only on providing ancillary services, the depth of discharge (DoD) of the ESS is reduced. This will help improve the operational lifetime of the ESS by mitigating the deep cycling. It will also improve the operational flexibility of the microgrid as seen by the controller by minimizing the amount of time of forced charging, that is, the times where the ESS cannot provide ancillary services because it must charge to replenish its stored energy.

In summary, a microgrid operating control scheme is presented that allows ESSs to focus on supplying only ancillary services rather than both energy and ancillary services. Including an ESS can reduce the total production cost of operating the microgrid. By allowing the ESS to focus mainly on providing ancillary services, namely regulation, the reliability of the microgrid in terms of the area control error can be improved. The lifetime of the ESS can also be extended by reducing the depth of discharge. Allowing the ESS to focus on ancillary services generally also reduces the amount of curtailed wind and solar generation since they have superior flexibility and can dispatch around available wind and solar generation in real time.

REFERENCES

[1] Kyung Soo K, McKenzie KJ, Yilu L, Atcitty S. A study on applications of energy storage for the wind power operation in power systems. In: Power engineering society general meeting. IEEE. Montreal: Canada; 2006.

[2] Wei L, Joos G. Performance comparison of aggregated and distributed energy storage systems in a wind farm for wind power fluctuation suppression. In: Power engineering society general meeting. IEEE. Tampa, FL: USA; 2007.

[3] Roebuck K. Energy storage: high-impact strategies – what you need to know: definitions, adoptions, impact, benefits, maturity, vendors; 2011.

[4] Feinberg EA, Genethliou D, editors. Load forecasting. Applied mathematics for restructured electric power systems. New York: Springer; 2005. p. 269−285.

[5] Liyan Q, Wei Q. Constant power control of DFIG wind turbines with supercapacitor energy storage. IEEE Trans Ind Appl 2011;47:359−67.

[6] Senjyu T, Datta M, Yona A, Funabashi T, Kim C-H. PV Output power fluctuations smoothing and optimum capacity of energy storage system for PV power generator. In: International conference on renewable energies and power quality (ICREPQ08); March 12−14, 2008.

[7] Alam MJE, Muttaqi KM, Sutanto D. A novel approach for ramp-rate control of solar PV using energy storage to mitigate output fluctuations caused by cloud passing. IEEE Trans Energy Conv 2014;29:507−18.

[8] Wroblewska M. Emergency generators – 10 second starting. ePOWER NEWS; 2011.

[9] Ling Q, Lu Y. An Integration of super capacitor storage research for improving low-voltage-ride-through in power grid with wind turbine. In: Presented at the power and energy engineering conference (APPEEC), Shanghai, China: Asia-Pacific; 2012.

[10] Thanh Hai N, Dong-Choon L. LVRT and power smoothening of DFIG-based wind turbine systems using energy storage devices. In: 2010 international conference on control auto-mation and systems (ICCAS), Gyeonggi-do: Korea; 2010. p. 1070−74.

[11] Iov F, Hansen AD, Sørensen PE, Cutululis NA. Mapping of grid faults and grid codes. Roskilde, Denmark: Risø National Laboratory; 2007.

[12] Raja ABA, Ajin S, Suja R, Williams AS. Review of low voltage ride through methods in PMSG. Inter J Res Sci Technol 2015;2 (1).

[13] Fu Q, Montoya LF, Solanki A, Nasiri A, Bhavaraju V, Abdallah T, et al. Microgrid genera-tion capacity design with renewables and energy storage addressing power quality and surety. IEEE Trans Smart Grid 2012;3:2019−27.

[14] Montoya LF, Fu Q, Nasiri A, Bhavaraju V, Yu D. Novel methodology to determine the optimal energy storage location in a microgrid and address power quality and stability, white paper: Eaton Corporation; 2014.

[15] Xisheng T, Wei D, Zhiping Q. Research on micro-grid voltage stability control based on supercapacitor energy storage. In: ICEMS 2009 international conference on electrical machines and systems, Tokyo: Japan; 2009.

[16] El-Deeb HM, Daoud MI, Elserougi A, Abdel-Khalik AS, Ahmed S, Massoud AM. Maximum power transfer of PV-fed inverter-based distributed generation with improved voltage regulation using flywheel energy storage systems. In: Industrial electronics society, IECON 2014 – 40th annual conference of the IEEE; 2014. p. 3135−41.

[17] Lazarewicz M.L., Rojas A. Grid frequency regulation by recycling electrical energy in flywheels. In: Power engineering society general meeting. IEEE, vol. 2, Denver; Colorado: 2004. p. 2038−42.

[18] Gayathri Nair S, Senroy N. Wind turbine with flywheel for improved power smoothening and LVRT. In: Power and energy society general meeting (PES). IEEE, Vancouver: BC, Canada; 2013.

[19] Seung-yong H, Woo-Seok K, Ji Hoon K, Chang-seop K, Song-yop H. Low speed FES with induction motor and generator. IEEE Trans Appl Supercond 2002;12:746−9.

[20] Eisenhaure DB, Kirtley Jr JL, Lesster LE. Uninterruptible power supply system using a slip-ring, wound-rotor-type induction machine and a method for flywheel energy storage. Google Patents; 2006.

[21] Akagi H, Sato H. Control and performance of a doubly-fed induction machine intended for a flywheel energy storage system. IEEE Trans Power Electron 2002;17:109−16.

[22] Wang L, Yu JY, Chen YT. Dynamic stability improvement of an integrated offshore wind and marine-current farm using a flywheel energy-storage system. Renew Power Gen 2011;5:387−96.

[23] Samineni S, Johnson BK, Hess HL, Law JD. Modeling and analysis of a flywheel energy storage system for Voltage sag correction. IEEE Trans Ind Appl 2006;42:42−52.

[24] Wang MQ, Gooi HB. Spinning reserve estimation in microgrids. IEEE Trans Power Syst 2011;26:1164−74.

[25] Kariuki KK, Allan RN. Evaluation of reliability worth and value of lost load. IEE Proc Gen Trans Distrib 1996;143:171−80.

[26] Krad I, Gao DW. Impact of PHEV on reserve scheduling: a MILP-SCUC method. In: North american power symposium (NAPS); September 2013. p. 1, 6, 22−24.

[27] Arroyo JM, Conejo AJ. Modeling of start-up and shut-down power trajectories of thermal units. IEEE Trans Power Syst 2004;19:1562−8.

[28] Perfect Power at the Illinois Institute of Technology, <http://www.iitmicrogrid.net/microgrid.aspx> [accessed 08.05.15].

CHAPTER 3

Interfacing Between an ESS and a Microgrid

3.1 INTRODUCTION

Power converter interface devices are important for using energy storage systems (ESSs) in microgrids. With the development of power electronics technology, many different power converters have emerged, which make the charging and discharging process of ESS controllable due to the bidirectional power flow ability of some converters.

Figure 3.1 shows some applications of power electronic converters for ESSs in microgrids, which may have a common DC and AC bus. The converter may differ for different ESSs, for example, a battery energy storage system (BESS) needs DC-DC or DC-AC converters to connect to a microgrid and for a flywheel energy storage system (FESS), a back-to-back converter is needed to perform AC-AC conversion; or a rectifier can be used to convert AC output power from FESS to DC to connect to a DC link in a microgrid.

In this chapter, several different power converters are discussed including DC-DC converters, AC-DC rectifiers, DC-AC inverters, and AC-AC converters. The structures and the principles of these converters are presented in the next sections.

In addition to converters, the management system is also essential for an ESS in a microgrid. The management system handles the operation of ESS, especially the charging and discharging process. In this chapter, the battery management system (BMS) is introduced. One major feature of the BMS is cell balancing of the battery package.

At the last part of this chapter, some applications of converters for ESS are presented with some specific control strategies.

3.2 DC-DC CONVERTER

DC-DC converters are used to change the voltage level at the DC source input to the desired DC output voltage level. There are several

Energy Storage for Sustainable Microgrid. DOI: http://dx.doi.org/10.1016/B978-0-12-803374-6.00003-2

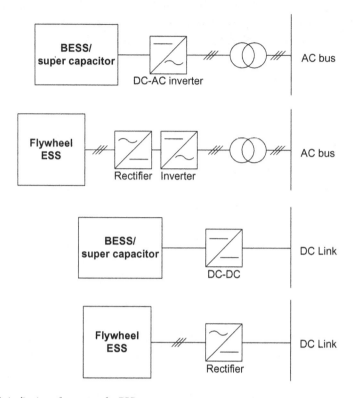

Figure 3.1 Applications of converters for ESS.

different types of DC-DC converters, namely step-down converter, step-up converter, and bidirectional DC-DC converter. A step-down converter converts a high voltage DC source into a low voltage output, while a step-up converter increases the DC source voltage to a higher output voltage. The bidirectional DC-DC converter can do both step-down and step-up conversions, and the power can flow in either direction between input and output.

Many ESSs are DC power sources, e.g., batteries and supercapacitors. Also, the DC link appears in microgrids in between back-to-back converters. So, a DC-DC converter can connect DC ESSs to the DC link in a microgrid. In addition, for a photovoltaic (PV) microgrid, the ESS can work to mitigate the intermittency of the PV output through control of DC-DC converter.

In this section, several DC-DC converters will be introduced. They cover the categories of step-down converter, step-up converter and bidirectional DC-DC converter.

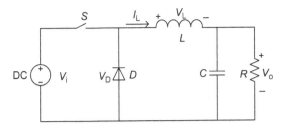

Figure 3.2 The circuit structure of buck converter.

3.2.1 Buck Converter (Step-Down Converter)

Buck converter is a simple and widely used voltage step-down device with high efficiency (e.g., 95% or above). Figure 3.2 shows a typical circuit of a buck converter.

In the buck converter, the inductor plays a major role to lower the input voltage. There are two states in the operation process of buck converter: the on-state and off-state of the switch S. During the on-state, the control signal closes the switch S. Since the source V_i is serially connected to the inductor and load, the current I_L through the inductor L is increasing. According to Faraday's law of induction, there will be a voltage V_L induced across the inductor. This opposing voltage V_L counteracts the voltage of the source and reduces the voltage on the load. At the same time, the inductor absorbs energy from the source and stores the energy in the form of a magnetic field.

On the other hand, if the switch S is opened by a control signal, this results in the off-state of the converter. In the off-state, the source voltage V_i is disconnected from the circuit by the switch. Due to the diode D, the current I_L through the inductor will continue to flow but its magnitude will drop. As a result, the induced voltage across the inductor will change its direction. Since there is energy stored in the inductor, the inductor becomes a source to supply the load by releasing its stored energy.

By switching between on-state and off-state constantly, the buck converter is able to decrease the voltage from the input to the output. If the current through the inductor never falls to zero during the whole process, the converter is said to be in continuous mode. Otherwise, it is in discontinuous mode. Figure 3.3 shows the current and voltage change during the continuous mode.

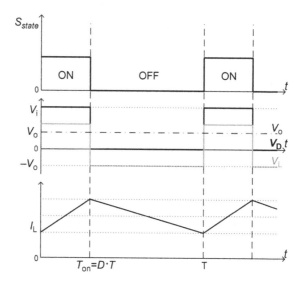

Figure 3.3 Voltage and current in continuous mode.

The voltage across the inductor is related to the change rate of its current, as given in Eq. (3.1).

$$V_L = L \frac{dI_L}{dt} \tag{3.1}$$

Therefore, in continuous mode, the change of current can be calculated for both on-state and off-state. Equation (3.2) denotes the current change in on-state.

$$\Delta I_{L_{on}} = \int_0^{t_{on}} \frac{V_L}{L} dt = \frac{V_i - V_o}{L} t_{on} \tag{3.2}$$

Similarly, for the off-state, the decrease of current through the inductor is computed by Eq. (3.3)

$$\Delta I_{L_{off}} = \int_{t_{on}}^{T} \frac{V_L}{L} dt = \frac{-V_o}{L}(T - t_{on}) = -\frac{V_o}{L} t_{off} \tag{3.3}$$

If the converter operates in a steady state, during a cycle, the current through the inductor at the beginning of the on-state will be the same as the current at the end of the off-state. This means that the accumulated current change during one operational cycle (i.e., one period consisting of one on-state and one off-state) is zero.

$$\Delta I_{L_{on}} + \Delta I_{L_{off}} = \frac{V_i - V_o}{L} t_{on} - \frac{V_o}{L} t_{off} = 0 \tag{3.4}$$

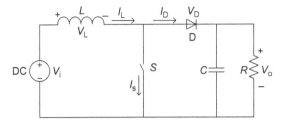

Figure 3.4 The circuit structure of boost converter.

Let $D = t_{on}/T$ be the switch duty cycle, and $0 < D < 1$. Equation (3.4) becomes:

$$\frac{V_i - V_o}{L} DT - \frac{V_o}{L}(1 - D)T = 0 \qquad (3.5)$$

$$\Rightarrow \frac{V_o}{V_i} = D \qquad (3.6)$$

So, by controlling the switch duty cycle of the converter, the output voltage V_o can be controlled. Also, as D is always blow 1, Eq. (3.6) shows that the output voltage is always lower than the input voltage.

3.2.2 Boost Converter (Step-Up Converter)

Like the buck converter, the boost converter has a simple structure, as shown in Figure 3.4. The function of a boost converter is to increase the input voltage to a higher output voltage. Again, the inductor in the circuit plays a major role to boost the input voltage.

The boost converter also has an on-state and an off-state. In the on-state, the switch S is in closed position, $I_D = 0$, the current I_L through the inductor L is increasing and the voltage across the inductor is equal to the source voltage. In this process, the inductor is storing energy from the source.

In the off-state, the switch is open. The load is reconnected to the source and the inductor. Hence, the current through the inductor will decrease, resulting in an induced voltage across inductor in the same direction of the source voltage. Now, the inductor becomes another source to supply the load by releasing its stored energy. Since the source and inductor are connected in series, the voltage on the load is

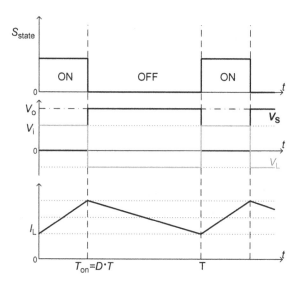

Figure 3.5 Voltage and current in continuous mode.

the aggregation of the voltage on inductor and source voltage, which means the output voltage is higher than the source.

$$V_o = V_L + V_i \qquad (3.7)$$

If the current through the inductor is always above zero, the converter is working in continuous mode. According to the relationship between the current and voltage of the inductor, the change of current during on-state and off-state can be obtained. In the on-state, only the source V_i influences the current through the inductor as Eq. (3.8) shows.

$$\Delta I_{L_{on}} = \int_0^{t_{on}} \frac{V_L}{L} dt = \frac{V_i}{L} t_{on} = \frac{V_i}{L} DT \qquad (3.8)$$

where D is the duty cycle. In the off-state, both the source and the load are connected with the inductor in series. The inductor current change is given in Eq. (3.9).

$$\Delta I_{L_{off}} = \int_{t_{on}}^{T} \frac{V_L}{L} dt = \frac{V_i - V_o}{L} (1 - D)T \qquad (3.9)$$

In the steady state, as Figure 3.5 shows, the current I_L at the beginning of on-state and at the end of the off-state will be the same, which means that the accumulated current change during a cycle is zero.

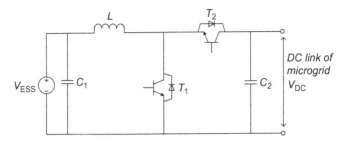

Figure 3.6 Bidirectional buck-boost converter.

Therefore, the following relationship of the current change through the inductor can be obtained:

$$\Delta I_{L_{on}} + \Delta I_{L_{off}} = \frac{V_i}{L}DT + \frac{V_i - V_o}{L}(1 - D)T = 0 \qquad (3.10)$$

which implies that

$$\frac{V_o}{V_i} = \frac{1}{(1 - D)} \qquad (3.11)$$

Because the duty cycle D is less than 1, the ratio in Eq. (3.11) will always be bigger than 1. This implies that the output voltage V_o is always higher than the input V_i. The gain of the boost converter can be controlled by the duty cycle D.

3.2.3 Bidirectional Buck-Boost Converter

In a microgrid system, charging and discharging of BESS requires the DC-DC converter to be bidirectional so that the power flow can change direction between the ESS and the microgrid. Typically, the voltage at the ESS side is lower than that at the microgrid side. That means that the ESS voltage must be stepped up so that the ESS can discharge to supply power to the microgrid. Conversely, when the ESS is being charged, the higher voltage at the DC-bus of the microgrid needs to be stepped down to the rated voltage of the ESS.

Figure 3.6 shows the structure of a bidirectional DC-DC converter. The transistors in the converter work as switches to connect or disconnect the circuit. This bidirectional DC-DC converter can be considered as a combination of a buck converter and a boost converter. In the buck converter mode, the transistor T_1 is always off, current flows from the DC-bus to the ESS source. By controlling the transistor T_2, the converter can decrease the microgrid side voltage V_{DC} to charge the ESS. If the converter works in boost converter mode, the transistor T_2 is always in its off-state and the diode on T_2 facilitates current flow in one direction from the ESS source to the microgrid. By controlling the duty cycle D of the transistor T_1, the converter is able to increase the output voltage V_{ESS} of the ESS so as to supply power to the microgrid.

A case study of bidirectional buck-boost converter will be given at the end of this chapter.

3.3 AC-DC AND DC-AC CONVERTER

AC-DC converters are also called rectifiers, which is a device to convert AC power into DC output. In a power system, three-phase rectifiers are very important for industrial applications and for the transmission of energy, e.g., in High Voltage Direct Current (HVDC) transmission system. In microgrids, the rectifier is useful to provide DC power to charge the ESS.

On the other hand, DC-AC converters are also called inverters, and are very important devices in power systems. For any DC power source, an inverter is needed to connect to an AC bus.

The structures of typical inverters and rectifiers are similar, which make it possible for a converter to function as either a rectifier or an inverter. For example, a voltage source converter (VSC) can be controlled to function as rectifier or inverter and allows bidirectional power flow.

In this section, different AC-DC rectifiers and DC-AC inverters will be introduced.

3.3.1 Single-Phase AC-DC Rectifier

In a single-phase rectifier, the input of the rectifier is one phase AC power. There are two major AC-DC rectifier types: half-wave rectifier and full-wave rectifier. The structures of these two types of rectifiers are very simple and no control signal is required.

3.3.1.1 Single-Phase Half-Wave Rectifier

This half-wave rectifier can have the simplest structure among all the AC-DC converters. As shown in Figure 3.7, only one diode D is needed in the half-wave rectifier. This diode limits the current flow in

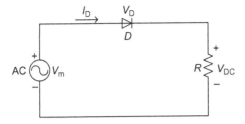

Figure 3.7 Single-phase half-wave rectifier.

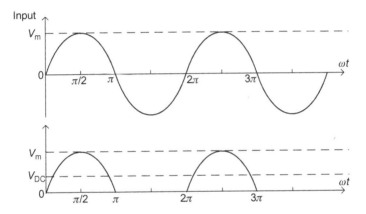

Figure 3.8 Waveforms of the input and output of half-wave rectifier.

one direction. This means that only half of the AC waveform can pass through the diode, as shown in Figure 3.8.

The input and output voltage waveforms of an example rectifier with a resistive load are shown in Figure 3.8. In the diagram, only the positive part of the input waveform does useful work. Thus the efficiency of the half-wave rectifier is very low.

According to the waveform in Figure 3.8, the voltage relationship of the input and output can be calculated. The average voltage on the load is:

$$V_{DC} = \frac{1}{2\pi} \int_0^{2\pi} V_m \sin \omega t \, d(\omega t) = \frac{1}{2\pi} \int_0^{\pi} V_m \sin \omega t \, d(\omega t)$$

$$\Rightarrow \quad V_{DC} = \frac{V_m}{\pi}$$

(3.12)

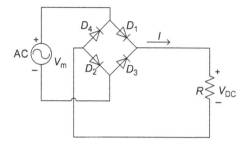

Figure 3.9 Single-phase full-wave rectifier.

where V_m denotes the amplitude of the input voltage. Further, the root-mean-square (rms) value of load voltage is:

$$V_{rms} = \sqrt{\frac{1}{2\pi} \int_0^{2\pi} (V_m \sin \omega t)^2 d(\omega t)} = \frac{V_m}{2} \tag{3.13}$$

Note that the output of the rectifier is half of the AC voltage source, which is not constant. Therefore, in many applications, a filter is needed to convert the half-wave DC voltage into a constant DC voltage.

3.3.1.2 Single-Phase Full-Wave Rectifier

Differing from the half-wave rectifier, the full-wave rectifier allows the input AC source to complete its current flow circuit in both positive and negative half of an AC cycle. The structure of the full-wave rectifier is shown in Figure 3.9.

There are four diodes in the full-wave rectifier circuit. When the AC source voltage is positive, the current flows through D_1 to the load and back to the AC source via D_2. When the AC source voltage is negative, the current flows via D_3-load-D_4 path. Either way, the current always goes across the load from the positive to the negative pole.

Figure 3.10 shows the waveforms of the input and output voltage of the full-wave rectifier. It can be considered as if rectifier flips the negative voltage to the positive side. Thus, the period of the waveform changes from 2π to π. Therefore, the average voltage across the load can be calculated as in Eq. (3.14).

$$V_{DC} = \frac{1}{\pi} \int_0^{\pi} V_m \sin \omega t \, d(\omega t) = \frac{2V_m}{\pi} \tag{3.14}$$

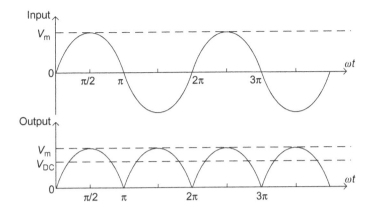

Figure 3.10 Waveforms of the input and output of full-wave rectifier.

And the rms value of load voltage is:

$$V_{rms} = \sqrt{\frac{1}{\pi} \int_0^\pi (V_m \sin \omega t)^2 d(\omega t)} = \frac{V_m}{\sqrt{2}} \qquad (3.15)$$

Compared with the half-wave rectifier, the average voltage for the full-wave rectifier is twice that of the half-wave rectifier's average voltage. The output rms voltage of full-wave rectifier is $\sqrt{2}$ times higher than that of half-wave rectifier.

3.3.2 Three-Phase AC-DC Rectifier

3.3.2.1 Three-Phase Half-Wave Rectifier

The star rectifier is a commonly used three-phase half-wave rectifier, which is like an aggregation of three single-phase half-wave rectifiers. The star rectifier can be more than three-phase by adding more diodes. Figure 3.11 shows the structure of a controllable three-phase half-wave rectifier. Each phase has a similar structure to one single-phase half-wave rectifier, and they are connected in parallel with each other. The diode's thyristors can be controlled to turn on or off at different times to produce different DC output voltages.

For the three-phase half-wave rectifier, if the control signal is always on for the thyristors, the three thyristors will work as normal diodes. Because the diode conducts only if the anode-to-cathode voltage is positive, the output voltage V_{DC} is in phase with highest voltage envelope, as shown in Figure 3.12.

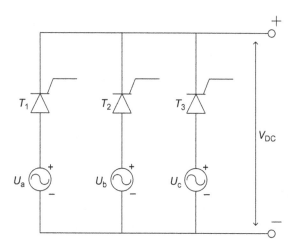

Figure 3.11 Controllable three-phase half-wave rectifier.

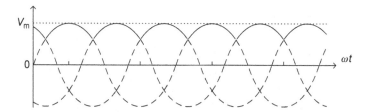

Figure 3.12 Half-wave rectification with diodes.

Besides the always-on mode, the control signal can make the thyristors conduct with a delay angle α for each phase. Assuming that the delay is counted from the crossing point between two phase voltages, the waveform of the output voltage for a resistive load looks like the form in Figure 3.13.

According to Figure 3.13, the output voltage has a period of $\frac{2}{3}\pi$. The average voltage on the resistive load can be calculated from:

$$V_{DC} = \frac{V_m}{\frac{2}{3}\pi} \int_{-\frac{\pi}{3}+\alpha}^{\frac{\pi}{3}+\alpha} \cos \omega t \, d(\omega t) = \frac{V_m \left(\sin \frac{\pi}{3}\right)}{\frac{\pi}{3}} \cos \alpha = \frac{3\sqrt{3}}{2\pi} V_m \cos \alpha$$

$$(3.16)$$

Because $V_m = \sqrt{2} V_{rms}$ for the input AC source voltage then,

$$V_{DC} = \frac{3\sqrt{3}}{2\pi} \sqrt{2} V_{rms} \cos \alpha \approx 1.17 V_{rms} \cos \alpha \qquad (3.17)$$

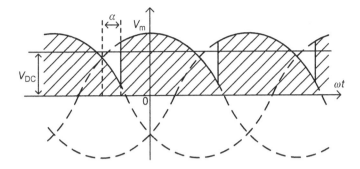

Figure 3.13 Waveform of the output of a controllable rectifier with delay α.

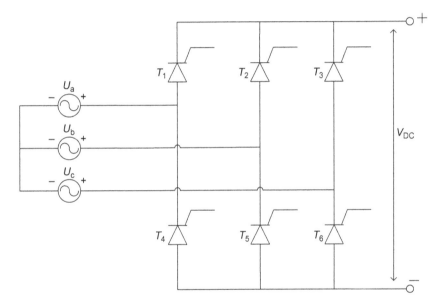

Figure 3.14 Fully controllable three-phase full-wave rectifier.

From Eq. (3.17), it is clear that the output of the star rectifier has the highest magnitude when the delay $\alpha = 0°$.

3.3.2.2 Three-Phase Full-Wave Rectifier

The three-phase full-wave rectifier has the bridge structure as shown in Figure 3.14. If all six thyristors are controllable, it is known as a fully controllable bridge converter. However, if the thyristors always conduct like diodes, the operation principle of the rectifier is similar to the single-phase full-wave rectifier discussed in preceding sections. This means that the negative half cycle voltage of each phase will flip to the positive side.

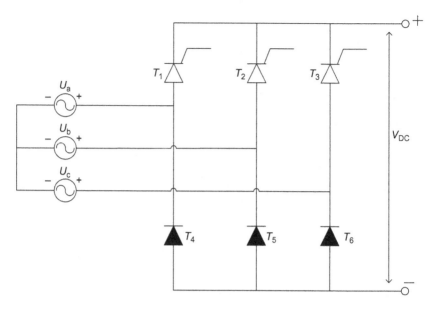

Figure 3.15 Half controlled three-phase full-wave rectifier.

By introducing the delay angle α to the control signals, the output voltage can be controlled like the three-phase half-wave rectifier. Equation (3.18) shows the calculation of the average voltage on the DC side.

$$V_{DC} = \frac{2V_m}{\frac{2}{3}\pi} \int_{-\frac{\pi}{3}+\alpha}^{\frac{\pi}{3}+\alpha} \cos \omega t \, d(\omega t) = \frac{3\sqrt{3}}{\pi} V_m \cos \alpha \qquad (3.18)$$

In a similar manner to the three-phase half-wave rectifier, the maximum output voltage happens when the delay $\alpha = 0°$.

For normal rectification applications, there is no need to control all the thyristors in the bridge. So the circuit can be simplified into a half controlled rectifier, as shown in Figure 3.15. In the half controlled bridge structure, the three thyristors, T_4, T_5, and T_6, can be replaced by diodes. By controlling the other three thyristors (T_1, T_2, and T_3), the rectifier can still achieve the desire performance. Since the diode is cheaper than thyristor, the half controlled bridge structure is cheaper.

3.3.3 Single-Phase DC-AC Inverter

The single-phase DC-AC inverter can convert DC input voltage into single-phase AC output voltage. Figure 3.16 shows the circuit of a

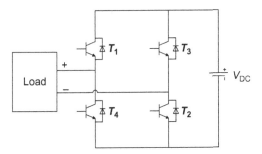

Figure 3.16 Single-phase DC-AC inverter.

single-phase DC-AC inverter. The circuit is an H-bridge structure with four controllable transistors.

A simple way to generate AC voltage on the load is called the Bipolar pulse width modulation (PWM) Technique [1]. In this method, the controller sends signal to turn on T_1 and T_2 during $0 < t < T/2$, which results in a positive current flow across the load. When $T/2 < t < T$, the controller turns off T_1 and T_2 and switches on T_3 and T_4, so the current through the load becomes negative. By switching between these two modes continuously, an AC voltage (square wave) with period T can be obtained across the load. In order to obtain sinusoidal AC output voltage, a filter is needed to remove the harmonics.

Note that T_1 and T_4 (or T_2 and T_3) should not conduct at the same time, otherwise the DC side will be shorted.

3.3.4 Voltage Source DC-AC Inverter
Today, most microgrids work in a three-phase AC power environment just like the traditional grid. In many applications, the ESS needs to connect to the AC bus in the microgrid. So a three-phase inverter is required. The structure of a three-phase inverter is similar to a controllable three-phase rectifier, thus many inverters are bidirectional and can work in DC-AC inverter or AC-DC rectifier mode.

The voltage source inverter (VSI) is a commonly used power inverter. It converts a DC voltage into a three-phase AC voltage. The structure of a three-phase VSI is illustrated in Figure 3.17. The three-phase VSI has six transistors to form a bridge structure with three legs.

The basic control scheme of VSI can be achieved by operating three transistor pairs coordinately, through a six-step sequencing control.

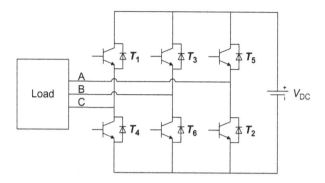

Figure 3.17 Voltage source DC-AC inverter.

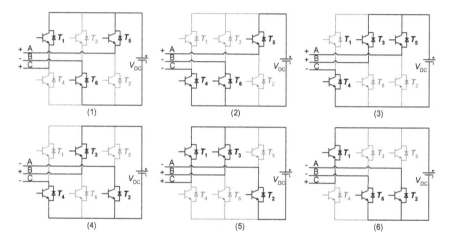

Figure 3.18 Six-step sequencing control of VSI.

The switches on the same branch (e.g., T_1 and T_4) are controlled by opposite signals so that they cannot conduct simultaneously, otherwise the DC source will be shorted by this branch. The six-step sequencing control is shown in Figure 3.18.

The control scheme contains six steps, and the duration of each step is $\pi/3$. At each step, the transistor pair on one branch switch their signals, but the other two branches do not change. The three-phase output voltage waveforms of the VSI with six-step sequencing control are shown in Figure 3.19.

There are many applications of VSIs in microgrids. One of the most important application is as a solar PV power inverter. Companies like Fronius, Outback and SMA, sell many types of PV inverters for

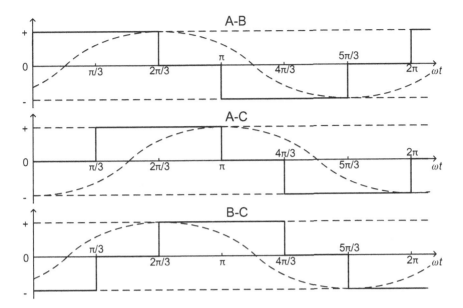

Figure 3.19 Waveform of six-step switching control.

Figure 3.20 PV inverter products. Copyright © SMA Solar Technology AG (Right), Copyright © OutBack Power Technologies (Bottom)

different ratings. In many of their products, the PV inverter not only converts the DC power from PV panels into AC output but also is equipped with some advanced features like Maximum Power Point Tracking (MPPT). Figure 3.20 shows PV inverter products from different companies.

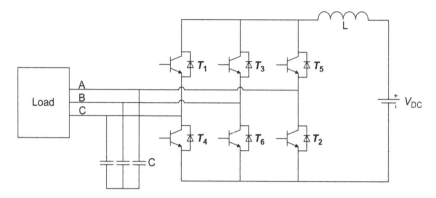

Figure 3.21 Current source DC-AC inverter.

3.3.5 Current Source DC-AC Inverter

The current source inverter (CSI) is another type of inverter. Different from VSI, the CSI converts DC current into an AC current. Figure 3.21 shows the structure of a CSI. It is clear that CSIs have similar structures to VSIs. But, in order to mitigate the high current variation in CSI, capacitors are usually installed on the AC side, while inductor is employed on the DC side.

3.4 AC-AC CONVERTER AND TRANSFORMER

AC-AC converters and transformers are also important converters for renewable microgrids and ESSs. In the renewable microgrid, some Distributed Generators (DGs) have AC power output, e.g., wind turbines and hydro generators. For ESS, the flywheel is an important AC storage system. Therefore, AC-AC converters and transformers can be used to control the output of renewables and the charging/discharging process of ESS. In addition, AC-AC converters and transformers can be employed to isolate the power source from the AC bus of the microgrid. For AC-AC converters, the DC link can separate the AC lines so that the effects of undesired harmonics caused by renewables can be reduced. By installing transformers, galvanic isolation can be achieved. Hence, the AC-AC converter and transformer can help to improve the stability of a microgrid.

3.4.1 Back-to-Back Converter

For renewables and ESS, a back-to-back structure is often used in the AC-AC converter. The charging and discharging of a FESS is usually

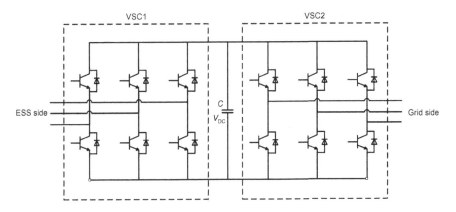

Figure 3.22 Back-to-back AC-AC converter.

controlled by a back-to-back converter. The back-to-back converter can control the input and output AC power with its transistors or Insulated-gate bipolar transistor (IGBT) switches. Figure 3.22 shows a typical back-to-back AC-AC converter, which is a combination of two VSCs with a DC link in between. So the back-to-back AC-AC converter can be controlled as two separate converters.

For example, in a FESS [2], the back-to-back converter absorbs energy from the AC bus to charge the FESS. The VSC2 on the grid side works as a rectifier to convert the AC power into stable DC power with constant voltage V_{DC} on the DC link. The controller for VSC2 can be designed as dual-loop controller with an inner current control and an outer voltage control loop. Meanwhile, the VSC1 works in inverter mode to convert the stabilized DC voltage into AC to supply the FESS. The VSC1 also has a dual-loop controller, which controls the current and speed of the flywheel's permanent magnet brushless DC machine (BLDCM).

In the discharging process, the VSC1 and VSC2 switch their operation modes. However, since the output of flywheel decreases when discharging, the VSC1 should be a boost rectifier with current-voltage dual-loop controller [3]. For VSC2, it works as an inverter to convert the DC power into stable AC power to supply the microgrid.

An application example of a back-to-back converter is the FESS based Power Control Module (PCM) of Beacon Power Company, as shown in Figure 3.23.

Representative flywheel energy storage module

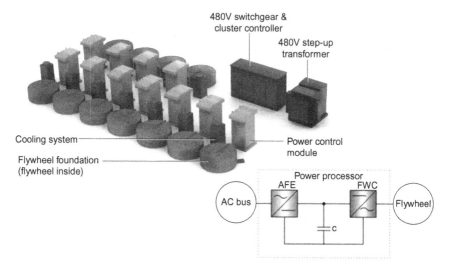

Figure 3.23 Beacon's Power Control Module [4].

The PCM can help the flywheel ESS to couple with the microgrid. The AFE (Active Front End) in the PCM can work as an active rectifier that converts line frequency AC power into DC, while the FWC (Flywheel Controller) of the converter inverts DC power into adjustable voltage and adjustable frequency AC for the flywheel. The PCM has been implemented in some projects, such as Energy Storage Pilot Project—Tyngsboro and Hazle Township project [4].

3.4.2 Transformer

Transformers are the most commonly used AC-AC converter device in power systems. But as opposed to power electronic converters, most transformers are uncontrollable, with a fixed conversion ratio. However, the transformer can be used in microgrids for galvanic isolation and to further step-up or step-down the AC voltage.

In Figure 3.24, example applications of transformers in microgrids with renewables and ESS are presented. It can be seen from the figure that the transformers can work with the inverters to interface with the AC bus. So the AC bus of the microgrid can be isolated from the renewables and ESS.

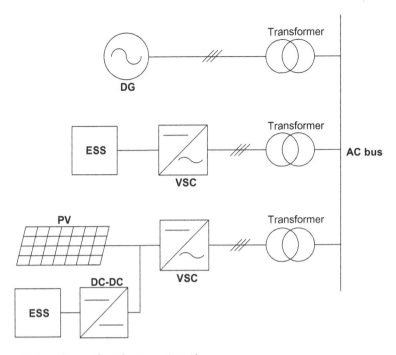

Figure 3.24 The application of transformers in microgrids.

3.5 BATTERY MANAGEMENT SYSTEM

BESS is one of the most applied energy storage technologies. Due to the fast response and high energy density of batteries, they have been widely applied in many microgrid projects.

In a BESS, hundreds of batteries work together to provide enough energy for the microgrid. Therefore, a BMS is required to coordinate the operation of all the batteries. The objective of the BMS is to provide protection, increase the life-span and maintain the stability of the batteries in the BESS. In order to accomplish the tasks, the BMS should have the functions of monitoring, computation, control and communication.

To protect the batteries from damage and maintain them in good condition, the BMS needs to monitor the temperature, voltage and current in the batteries. Overheating the battery is very dangerous, and may lead to explosion and fire. Hence, temperature monitoring is very important to ensure that the batteries work in stable conditions. The voltage information of the battery indicates its state, which is useful to identify the state-of-charge (SOC) of the batteries, so that overcharging or over-discharging can be prevented. Furthermore, the current of batteries should stay in a safe range, especially during the charging process. If the charging current is too high, it will not only damage the battery's health, but also causes accidents. However, if the charging current is too low, it will take a very long time to fully charge the battery.

The BMS also needs to perform many computations. For example, the SOC and state-of-health of the batteries should be calculated based on the voltage and current information.

In addition to charging and discharging control process of each battery, the SOC of each battery needs to be balanced. For charging and discharging control, the BMS should cooperate with the DC-DC converter to have the desired charging current.

For many control strategies, operation of the BESS is coordinated with other sources to achieve the optimal operation of the microgrid. So, the BMS needs to communicate with other devices in the microgrid. In a microgrid with multi-level control structure, the BMS will communicate with the central controller to report the conditions of batteries and receive control tasks.

There are two major topologies of BMS: centralized BMS and master-slave BMS. In this section, these two architectures will be introduced. Then the cell balancing technology will be presented.

3.5.1 Centralized BMS
Centralized BMS architecture has one central BMS in the BESS. All the battery packages are connected to the central BMS directly. The structure of a centralized BMS is shown in Figure 3.25.

The centralized BMS has some advantages. First, it is more compact. Second, the centralized BMS solution is the most economical since there is only one BMS. However, the disadvantages of centralized BMS are obvious. As all the batteries should be connected to the BMS directly, the BMS needs a lot of ports to connect with all the battery packages. However, there are too many wires in the BESS, making maintenance difficult. Therefore, if the BESS contains a lot of battery packages, a centralized BMS is not a good option.

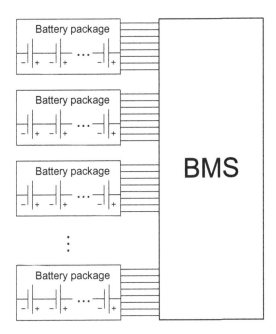

Figure 3.25 Centralized BMS.

3.5.2 Master-Slave BMS

The master-slave BMS has a different structure from the centralized BMS, as shown in Figure 3.26. In the master-slave BMS structure, there are two different types of BMS: the master BMS and the slave BMS.

For each battery package in the BESS, there is a slave BMS. The slave BMS controls the charging and discharging processes and monitors the conditions of each battery package. Also, it performs the cell balancing feature for its battery package. The master BMS assigns the tasks for each battery package and determines which battery packages should work. Moreover, the master BMS deals with communication between the BESS and the microgrid controller.

3.5.3 Cell Balancing

In each battery package, several battery cells are connected in series to achieve a bigger capacity. However, the quality or ability of each battery cell is not identical, which may result in imbalance among cells during charging or discharging.

The imbalance among battery cells will decrease the available energy of the overall battery package. As illustrated in Figure 3.27a, in

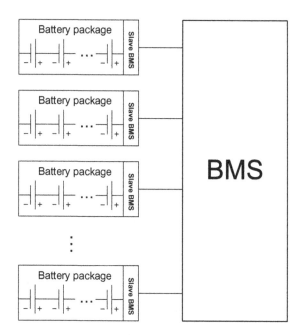

Figure 3.26 Master-slave BMS.

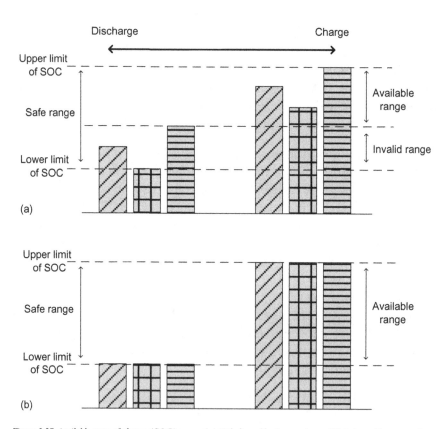

Figure 3.27 Available state-of-charge (SOC) range. (a) Unbalanced battery package. (b) Balanced battery package.

which three cells are connected in series to form a battery package, there are two unbalanced states in which each cell has a different SOC. Due to the safe range requirement of the battery's SOC, the battery package stops charging or discharging when one of the cells reaches the limits. Therefore, for the unbalanced situation, the battery package will not continue to charge or discharge when the battery cell with the highest SOC reaches the upper limit or the battery cell with lowest SOC reaches the lower limit. As a result, the cells within the battery package cannot be fully charged or fully discharged, except the battery cell with highest SOC or lowest SOC. Thus, the available range of SOC for charging or discharging is reduced. If the batteries are balanced, as shown in Figure 3.27b, all the battery cells can be charged and discharged fully, and the available range of battery package is maximized.

On the other hand, the parallel-connected battery cells can self-balance, so cell balancing is not needed in this structure. There are two different cell balancing techniques, passive cell balancing and active cell balancing.

3.5.3.1 Passive Cell Balancing
The commonly used passive cell balancing approach is known as "resistor bleeding balancing." It uses a bypass circuits to balance each battery cell. The circuit for passive cell balancing is shown in Figure 3.28.

There are some field-effect transistors (FET) and resistors connected to each cell in parallel. The balancing process can happen during both charging and discharging periods. In the charging process, the BMS calculates the SOC for each battery cell and finds out the highest one (suppose it is B_i). Then, the BMS sends a signal to turn on S_{ia} FET, and this will cause a current to flow through the resistor R_{ia}. This current creates a voltage on resistor R_{ia} to turn on the FET S_{ib}. As a result, the charging current will not flow in battery B_i and bypass through the resistor R_{ib}. For the discharging process, the BMS will also check the battery with highest SOC (B_i). By turning on S_{ia}, the FET S_{ib} will be turned on due to the same reason in charging process. So, the battery B_i will start to discharge as it is connected to resistor R_{ib}.

After performing the balancing process for a while, all the battery cells in the battery package will have the same SOC. However, as the current is bypassed through the resistors, it is obvious that passive

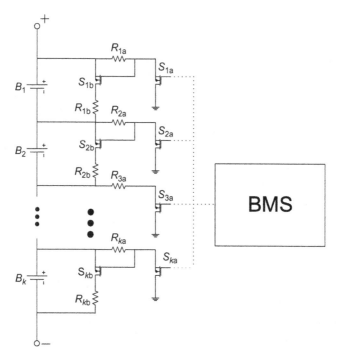

Figure 3.28 Passive cell balancing BMS.

cell balancing is not energy efficient. The energy is wasted as heat on the resistors. Hence, passive cell balancing is not an efficient way to manage the batteries.

3.5.3.2 Active Cell Balancing

Time-shared cell balancing is a popular method to balance the battery cells in BESS. The circuit of time-shared active cell balancing is shown in Figure 3.29. BMS obtains the battery voltage through the cell selection switches (i.e., $S_{2,1}$ and $S_{2,2}$ for battery B_2) and ADC (analog-digital converter) switch (S_a and S_b). Because the series-connected battery cells are at different voltage reference levels, the battery voltage can be obtained by a flying capacitor.

The operation process is as follows. When the BESS is at discharging mode, FETs S_i and S_{i+1} makes a connection between one of the battery cells and the flying capacitor C_1, and the battery voltage is transferred to the flying capacitor. Hence, the connection status is changed from (S_i, S_{i+1}) to (S_a, S_b) and the BMS reads the voltage of the flying capacitor. These switches are sequentially turned on from the first battery cell to the last one. Therefore, by calculating the SOC

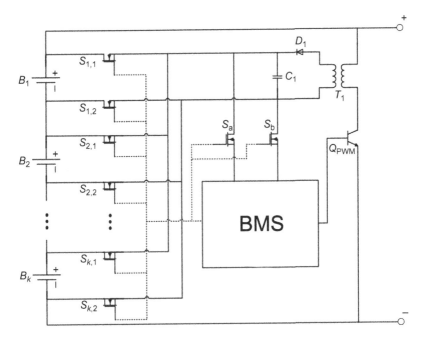

Figure 3.29 Active cell balancing BMS.

with this voltage information, the battery cell with the lowest SOC can be located. In this state, the PWM switch (Q_{PWM}) for DC/DC converter is not operated.

Suppose the battery with the lowest SOC is B_i. When battery cells are being charged from the microgrid, FETs $S_{i,1}$ and $S_{i,2}$ connect the battery B_i to the transformer T_1. The BMS controls the FET Q_{PWM} with PWM signal to transfer the current from input through the transformer T_1. So the battery cell B_i can be charged directly by the transformer. This process is continued until all the battery cells have the same SOC and thus the cell balancing is achieved.

3.6 APPLICATIONS OF CONVERTERS FOR ESS IN MICROGRID

In this section, several applications of converters for ESS will be introduced including the application of DC-DC converter to integrate ESS with renewables and the *dq* decoupled control for inverter.

3.6.1 ESS Integration with Renewables

Currently, renewable energy generation is the major power source of many microgrids. Since the outputs of many renewable energy sources

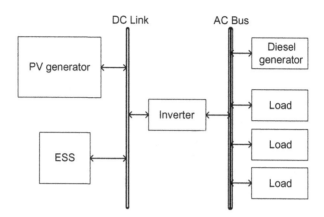

Figure 3.30 Connect ESS to the DC link.

are intermittent, ESS can work with renewables to improve their power quality. In many applications, the ESS is connected to the DC link, as shown in Figure 3.30. The DC link is a bus located behind the inverter, which is connected to the AC bus of the microgrid. The DC link is important to isolate the power source from the grid and improve the power quality since the output of generator may contain undesired harmonics [5].

Suppose that the ESS output is DC power, e.g., from BESS, the bidirectional buck-boost converter can be installed within the ESS in order to charge and discharge the ESS. If the output of ESS is AC, e.g., FESS, it can be rectified into DC power at first, and then connected to the bidirectional DC-DC converter. In Figure 3.31, these two types of structures are illustrated. Note that the rectifier connecting the FESS to the DC-DC converter should be bidirectional, which can be implemented with a VSC.

The function of the bidirectional DC-DC converter is to control the charging and discharging process of the ESS. Proportional-integral (PI) control is the common control strategy for this application. The controller should generate appropriate gate signal for the power electronic switches in the bidirectional DC-DC converter so that the output voltage or input current is able to arrive at desired value. An example controller structure is presented in Figure 3.32.

When the ESS is discharging, the control system changes S_1 to the off position and S_2 to the upper position to activate the discharging

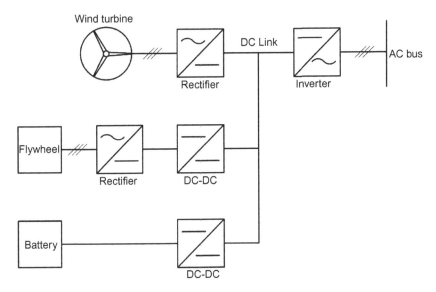

Figure 3.31 Configuration of AC output ESS and DC output ESS.

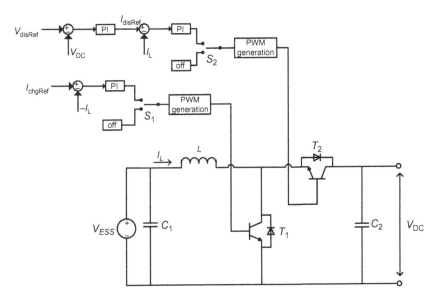

Figure 3.32 Controller design of the bidirectional DC-DC converter.

controller. Hence, the bidirectional DC-DC converter is working in boost converter mode to increase the output voltage of ESS to match the voltage on DC link. The discharging controller is a voltage-current dual-loop controller; the inner loop is a current loop to stabilize the

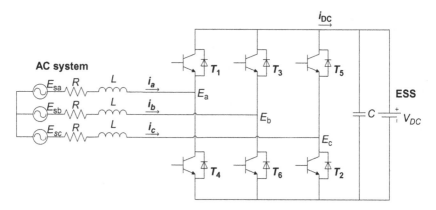

Figure 3.33 The model of grid-tied inverter.

discharging current through the inductor L, while the outer voltage loop controls the output voltage of the converter. The output of the outer controller is connected to a PWM generation module, which converts the control signal into the PWM gate signal for the transistor T_2.

In the charging process, the control system changes the position of switch S_1 to connect to the charging controller and turns S_2 to the off position. Thus, the converter works in buck converter mode and it decreases the voltage on the DC link to charge the ESS with desired current. If the charging current is too high or too low, it is bad for the health of the battery. Therefore, the reference of the charging controller, I_{chgRef}, is the desired charging current. The error between the reference and the actual charging current is input to a PI controller. Like in the discharging controller, a PWM generation module converts the control signal from the PI controller into a PWM gate signal and sends it to the transistor T_1.

3.6.2 *dq* Decoupling Control for Grid-Tied Inverter

For many storage systems or DGs, the grid-tied inverter is the last converter to interface with the grid. It converts the DC power from DC ESS or DC link to AC power to supply the microgrid. Also, it absorbs AC power from the microgrid to charge the ESS. Most grid-tied inverters are VSI. The structure of a VSI is shown in Figure 3.33.

The basic control purpose is to control the real power and reactive power output, through decoupled PQ control. The direct-quadrature (dq) decoupling control is to control the power output of inverter by converting the *abc* three-phase power into *dq*-axes so that the real power and reactive power can be regulated independently.

According to Figure 3.33, the relationship between the current and voltage of each phase is shown by Eq. (3.19).

$$\begin{cases} L\dfrac{di_a}{dt} + Ri_a = E_{sa} - E_a \\[2mm] L\dfrac{di_b}{dt} + Ri_b = E_{sb} - E_b \\[2mm] L\dfrac{di_c}{dt} + Ri_c = E_{sc} - E_c \end{cases} \tag{3.19}$$

By performing the *dq* reference frame transformation, the current-voltage equation in *dq*-axes is obtained as in Eq. (3.20).

$$\begin{cases} L\dfrac{di_d}{dt} = -Ri_d + \omega Li_q + V_{sd} - V_d \\[2mm] L\dfrac{di_q}{dt} = -Ri_q - \omega Li_d + V_{sq} - V_q \end{cases} \tag{3.20}$$

where i_d and V_d denote the output current and voltage on *d*-axis. i_q and V_q represent the values on *q*-axis. Now, assuming that

$$\begin{cases} v'_d = L\dfrac{di_d}{dt} + Ri_d \\[2mm] v'_q = L\dfrac{di_q}{dt} + Ri_q \end{cases} \tag{3.21}$$

Then, since v'_d and v'_d contains the first-order derivative of i_d and i_q, Eq. (3.20) becomes

$$\begin{cases} V_d = V_{sd} - v'_d + \omega Li_q \\ V_q = V_{sq} - v'_d - \omega Li_d \end{cases} \tag{3.22}$$

v'_d and v'_d can be obtained by PI controller as in

$$\begin{cases} v'_d = k_{p1}\Delta i_d + k_{i1}\int \Delta i_d dt \\ v'_q = k_{p2}\Delta i_q + k_{i2}\int \Delta i_q dt \end{cases} \tag{3.23}$$

where $\Delta i_d = (i_{dref} - i_d)$. So the decoupled output current i_d and i_q of the inverter can be controlled by performing the control strategy given

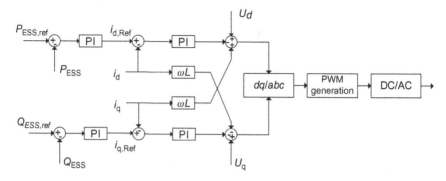

Figure 3.34 PQ controller structure for inverter.

by Eq. (3.22). The calculation results V_d and V_q can be transformed back to *abc*-axis to control the inverter. In addition, by employing another PI controller, output current references i_{dref} and i_{qref} are generated from the real power and reactive power output errors. Therefore, the structure of the PQ controller for the grid-tied inverter can be designed as in Figure 3.34.

As shown in Figure 3.34, a second PI controller is used to output the decoupled voltage signal in *d*-axis and *q*-axis respectively. The results are compensated by $\omega L i_q$ and $\omega L i_d$, respectively. After that, the control signal is transformed into *abc*-axis and then converted into a PWM signal to control the inverter.

3.7 CASE STUDIES

3.7.1 The Control of Bidirectional Buck-Boost Converter

In recent years, with the development of battery technology, the performance of battery has increased dramatically while at the same time the price has gone down. So BESSs are used more often in the new microgrid projects. In this section, a case study of bidirectional buck-boost converter is presented as a simulation model in Simulink.

As Figure 3.35 shows, the converter has two IGBTs to deal with the buck mode and boost mode, respectively. The IGBTs are controlled by a controller. The controller of the converter has five inputs, which are "mode, I_{ref}, V_{ref}, I_{fb}, and V_{fd}." The "mode" determines the operation mode of converter. If the input of "mode" is 1, the converter works in buck mode and the BESS is being charged, while the converter is in

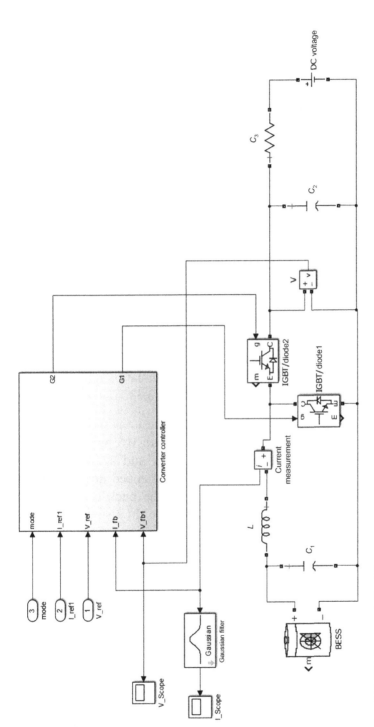

Figure 3.35 DC-DC converter in Simulink.

boost mode to discharge the BESS if "mode" is 0. I_{ref} and V_{ref} are the reference of charging current and the reference of discharging voltage. I_{fb} and V_{fd} are the inductor current and the converter voltage feedbacks. The output of a converter connects to a DC voltage source and a resistor, which represents a DC link in a microgrid.

The structure of the controller is shown in Figure 3.36. It consists of two proportional-integral-derivative (PID) controllers, one for the buck converter and the other for the boost converter. Each controller has a saturation to limit the output of PID to within -0.95 and 0.95 so that the control signal can be a valid duty cycle signal for the IGBTs. The PWM generators convert the duty cycle signal into a PWM signal to operate the IGBTs. When the "mode" signal is 1, the IGBT2, which connects to the G2 port in Figure 3.35, is linked to the buck controller. And the IGBT1 (G1 port) is always in the off-state due to the constant input 0. So the system becomes a buck converter and the DC voltage source can charge the BESS with the desired current value. If the "mode" signal is set to 0, the controller switches to boost mode, in which the IGBT1 is controlled by the boost controller and the input of IGBT2 is set to constant 0. In this mode, the converter can output energy from BESS to the DC link with desired voltage value. The simulation results of these two control modes are shown in Figures 3.37 and 3.38.

In Figure 3.37, the charging current control result is shown for 60 s. At first, the charging current reference is set to 0.5 A. At 15 s, the charging current reference is changed to 0.8 A and again to 1 A at 30 s. Finally, the charging current reference is stepped down to 0.6 A at 45 s. This simulation result indicates that the buck converter can control the charging current as the desired value. In a similar fashion, the result in Figure 3.38 shows that the voltage can follow the reference value well during ESS charging mode.

3.7.2 *dq* Control of VSI

In order to verify the validity of *dq* decoupling control in a grid-tied inverter system, a modeling and simulation case has been established in a Matlab/Simulink platform. The modeled grid-tied inverter system is presented in Figure 3.39.

As shown in Figure 3.39, the DC voltage source, which can be viewed as the DG in microgrid, is a battery. In this case, the battery will either provide power for the microgrid during discharging mode or

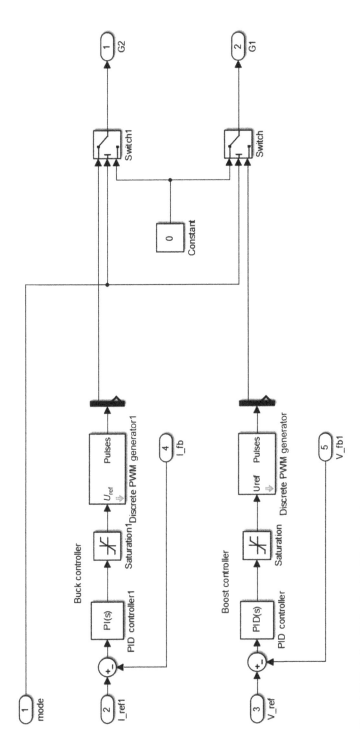

Figure 3.36 Controller of bidirectional buck-boost converter.

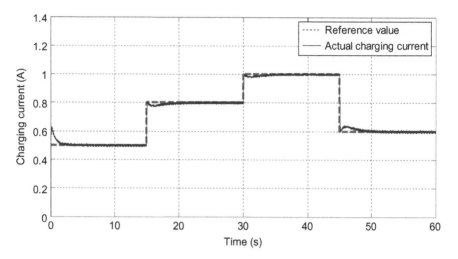

Figure 3.37 Charging process.

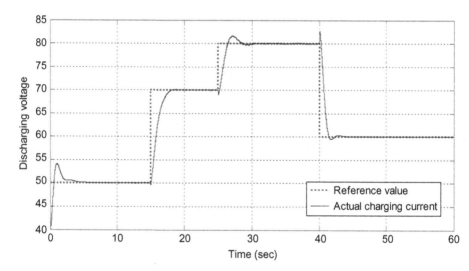

Figure 3.38 Discharging process.

extract power from microgrid during charging mode. Its working mode depends on reference signal of the system. Along with the battery, a bidirectional VSC (IGBT converter) can either convert DC power into AC power like a conventional inverter or regulate AC power into DC power as a rectifier. An inductor-capacitor-inductor (LCL) filter is necessary for removing high order harmonics to avoid power quality problems in the microgrid. For the ESS system in Figure 3.39, the output voltage of battery is low and thus AC voltage

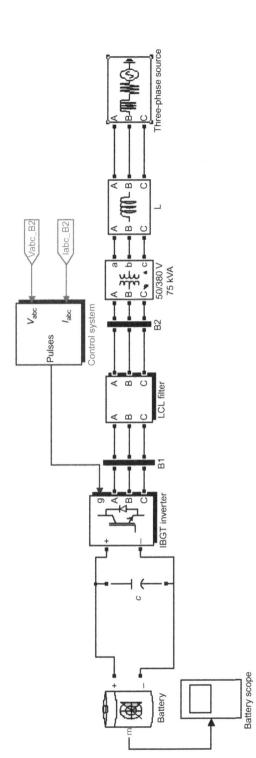

Figure 3.39 Grid-tied inverter system in Simulink.

Table 3.1 Battery Parameters			
Battery Type	Nominal Voltage	Rated Capacity	Initial SOC
Nickel-metal-hybrid	100 V	6.5 Ah	50%

Table 3.2 Main Circuit Parameters					
Inductors of LCL Filter	Capacitor of LCL Filter	Voltage Amplitude of Three-Phase Voltage Source	Voltage Source Resistance	Voltage Source Inductance
500 μH100 μH	20 μF	311 V	0.1 Ω	0.1e − 3H

output is not high enough for the microgrid. Therefore a transformer is used to change the voltage level. B1 and B2 are measurement tools which can detect three-phase voltage and current information. The control system is responsible for the control of the VSC. Some parameters of the VSC system have been listed in Tables 3.1 and 3.2. A detailed control system diagram is given in Figure 3.40.

Note that in the PQ control block, if the real power reference is positive, the ESS provides real power for microgrid through the VSC; and if the reactive power reference is positive, the ESS provides reactive power for microgrid through the VSC. However, if the real power reference or reactive power reference is negative, the power flow direction will be opposite.

The feedback signals are three-phase voltage and three-phase current from LCL output. The measured three-phase voltage is abbreviated as V_{abc} and three-phase current as I_{abc}. Then V_{abc} and I_{abc} are converted into the dq-axis for decoupling control. V_{abc} is converted into V_d and V_q while I_{abc} is converted into I_d and I_q. Whereas, a phase-lock-loop (PLL) is included in order to synchronize the output of VSC with the grid system. Real power and reactive power are calculated in PQ computation block based on values of V_d, V_q, I_d and I_q. Real power and reactive power are fed into PQ control block, as shown in Figure 3.41.

In order to control real power and reactive power independently, a dual-loop control strategy is employed in this case. From Figure 3.41, I_d reference signal and I_q reference signal for current regulation are generated from the PQ control block. These two reference signals, I_{dref} and I_{qref}, are fed into current regulator, as illustrated in Figure 3.42.

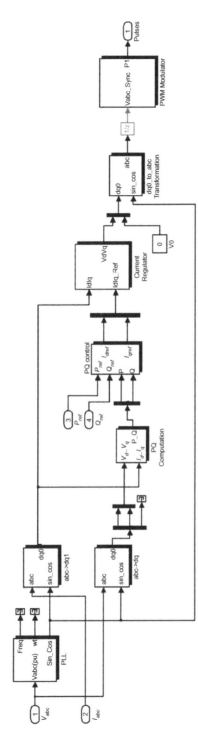

Figure 3.40 Control system diagram.

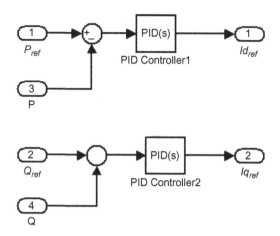

Figure 3.41 PQ control block.

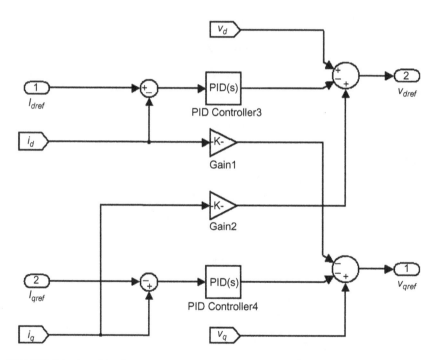

Figure 3.42 Current regulator diagram.

The establishment of current control structure follows Eqs. (3.20)–(3.23) in Section 3.6.2. Gain1 and Gain2 represent the coefficient ωL. The output signals V_{dref} and V_{qref} of the current regulators are used to generate PWM driving signals.

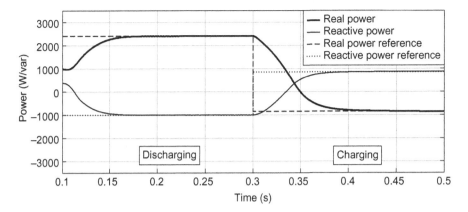

Figure 3.43 Output real power and reactive power.

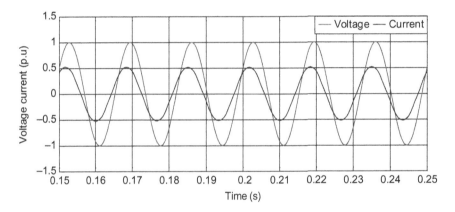

Figure 3.44 Output voltage and current in phase A.

First, the real power reference value is set at 2400 W and the reactive power reference value at −1000 var. VSC system simulation results are given in Figures 3.43–3.46. It is worth mentioning that the simulation time range of all figures (except Figure 3.44) is 0.5 s.

Figure 3.43 presents the real power and reactive power simulation results of grid-ties inverter system. The thicker solid line represents real power curve and the thinner solid line represents the reactive power curve. Also dotted line indicates real power reference and dashed line indicates reactive power reference. In this example, step signals are introduced into both the real power reference and the reactive power reference. Specifically, first the real power reference value is changed

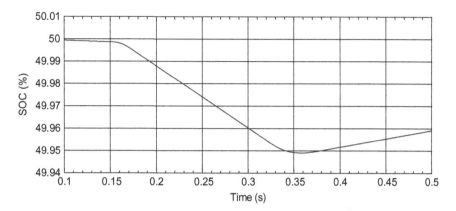

Figure 3.45 Battery SOC curve when discharging.

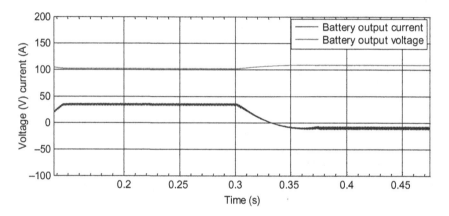

Figure 3.46 Battery voltage and current waveform during charging mode.

from 2400 to −1000 W and reactive power reference is changed from −1000 to 1000 var at 0.3 s. From Figure 3.43 it can be seen that the rise time of grid-tied inverter control system is less than 0.2 s and real power and reactive power are precisely controlled to the reference values. The simulation results are divided into two different sections which are battery charging section and battery discharging section. In conclusion, established VSC system is either able to charge the battery or provide power for microgrid.

Figure 3.44 shows the VSC output voltage and current waveform in phase A. It can be seen that voltage and current waveform are not in phase since reactive power output is not zero during 0.15−0.3 s.

Figure 3.45 presents battery SOC status of VSC system. Since simulation time is short, SOC does not change much from its initial value of 50%. Figure 3.45 presents the SOC curve during the discharging and charging process. This zoomed figure shows that power is indeed provided from battery to microgrid at discharging working mode and power is extracted from microgrid to battery at charging working mode.

Figure 3.46 shows battery voltage and current status of VSC system during discharging and charging working mode. From Figure 3.46 it can be seen that battery current was changed from positive to negative because battery working status is switched from discharging to charging.

REFERENCES

[1] Rashid M. In: Power electronics handbook. 3rd ed; 2011, Butterworth-Heinemann, Oxford, UK.

[2] Song X, Haifeng W. Simulation and analysis of back-to-back PWM converter for flywheel energy storage system. In: Electrical machines and systems (ICEMS), 15th International conference on 2012. p. 1—5.

[3] Junxing Z, Xinjian J, Lipei H. A novel dynamic voltage restorer with flywheel energy storage system. In: Electrical machines and systems, ICEMS 2008. International conference on 2008. p. 1995—9.

[4] System installation. Available from: <http://beaconpower.com/system-installation/>.

[5] Carlsson A. The back to back converter — control and design. Department of Industrial Electrical Engineering and Automation, Lund Institute of Technology, Lund, Sweden; 1998.

Coordinated Frequency Regulation of BESS with Renewable Generation in Microgrid

Note: This chapter is available on the companion website: http://book-site.elsevier.com/ 9780128033746.

Energy Storage for Sustainable Microgrid. DOI: http://dx.doi.org/10.1016/B978-0-12-803374-6.00012-3

Sizing of Energy Storage Systems for Microgrids

5.1 INTRODUCTION

The importance of ESS sizing problems stems from both economical and operational reasons. The investment cost of the ESS is dependent on its power rating and energy rating. Oversized ESS introducs high capital cost to the microgrid whereas undersized ESS may not be able to provide the desired economic or operational benefits [1-3]. Moreover, the integration of an improperly sized ESS will not be helpful for frequency regulation in islanded microgrids [4]. Thus, it is crucial to determine the suitable size for the ESS. Sizing ESS in a microgrid can be categorized as an optimization problem for which different optimization methods have been used. Among those methods are Particle Swarm Optimization (PSO) [4], Linear Programming [1,2], Genetic Algorithm (GA) and Dynamic Programming (DP) [5]. The optimization problem solution depends on its objective which in turn depends on the ESS application. In islanded microgrids, ESS is normally implemented to provide frequency and voltage regulation support since a strong main grid is not available for such function. Thus, the ESS should help maintain the power balance in the microgrid system. For grid-connected microgrids, however, the objective of the optimization problem mainly focuses on minimizing the total cost including the microgrid operation cost and the ESS investment cost. Since the energy can be transferred to and from the main grid, the ESS can be used to perform energy arbitrage application in grid-connected microgrids. This is generally done by storing energy in low price periods and discharging this energy in high price periods to make a profit. This will be shown later in Section 5.3 case study to explain the importance of finding the appropriate size for the ESS in microgrid applications. A brief review of some of the existing ESS sizing methods is given in Section 5.2.

Energy Storage for Sustainable Microgrid. DOI: http://dx.doi.org/10.1016/B978-0-12-803374-6.00005-6

5.2 ESS OPTIMAL SIZING METHODS

There are various energy storage sizing methods available to optimally design and select different energy storage technologies. This section focuses on discussing some of these techniques/methods. As every individual microgrid has different energy source components and capacity, it is imperative to study individual application case for optimal ESS and its sizing. Both energy rating and power rating should be considered while making the decision. Discussing all the available microgrid cases for energy storage sizing is beyond the scope of this book, however, a few cases from recently published papers are summarized to give the readers a comprehensive overview.

In [1] and [2], for example, mixed integer programming (MIP) is used to formulate the ESS sizing problem in grid-connected microgrid with the objective of minimizing the microgrid total cost. The objective function is as follows:

$$Min(IC + MC + OC) \qquad (5.1)$$

where,

IC = ESS investment cost
MC = ESS maintenance cost
OC = Microgrid operating cost

The following aspects should be considered to define system constraints:

1. The load should be satisfied for every hour; i.e., the power available from the main grid and the ESS should be sufficient.
2. The units should meet the reserve requirement.
3. The total amount of fuel that is to be consumed should be within the scheduling limits.
4. Emission by the units is also constrained to a certain limit.

Apart from the system constraints, storage constraints and unit constraints are also defined during problem formulation. Unit constraints usually consist of ramping up and down limits, unit minimum up time and down time limit.

Using these objective functions and constraints, expansion planning cost can be obtained for both ESS rated power and energy. It can be observed that as the storage rated power increases, so does the investment and maintenance cost of ESS while the operating cost of the microgrid decreases. Using these two curves, a trade-off can be obtained and the optimal ESS size can be determined. This method is explained with more details in the next section as it is implemented to determine the optimal ESS size in a case study.

In [2], however, the problem is extended by introducing reliability constraints. This is achieved by implementing a stochastic technique generating microgrid operation scenarios. The state of each component within the microgrid and the generation of renewable energy resources are obtained in each scenario. Since the number of generated scenarios is expected to be large, a scenario reduction technique is utilized to reduce the computational burden. This, of course, has an impact on the solution accuracy. Thus, the trade-off between the problem solution accuracy and the computational burden must be performed. In order to examine the reliability of the system, a loss of load expectation (LOLE) index is used.

Reference [6] studies the impact of different ESS technologies and sizes on the operation cost of a stand-alone microgrid. The modeled microgrid comprises a diesel generator, a wind turbine, and a dumb load, which is required to absorb any surplus energy from the microgrid generation. For each ESS technology, the optimal size that yields the lowest energy price is determined and compared to other technologies. A knowledge-based expert system (KBES) controller is used to schedule the diesel generator output power as well as the ESS charging/discharging cycles. Moreover, the controller is responsible for determining the amount of power that will be consumed by the dumb load. The studied ESS technologies include lead-acid batteries, nickel-cadmium batteries, sodium-sulfur batteries, zinc-bromine batteries, vanadium redox batteries (VRB), sodium-polysulfide batteries, superconducting magnetic energy storage (SMES), flywheel energy storage, electrochemical capacitors and large compressed air energy storage (CAES). These technologies are characterized by their power rating and energy rating, average life span, and round trip efficiency. An iterative method is used to find the energy rating and power rating of each technology that yield the lowest energy price. It is found that Lead-Acid battery is the best option for this application. In general, it is concluded that long-term energy storages give lower average price per KWh than short-term energy storages for the studied microgrid model.

In [7], an algorithm is proposed to size a battery energy storage system (BESS) in islanded microgrid. The objective of the problem is to minimize the cost of one-day unit commitment. In addition to the typical unit commitment problem cost which normally includes the conventional generators' generation cost and start up cost, a spinning reserve cost is considered. The objective function is defined as:

$$\min\left(\sum_t \sum_{i \in G} r_n R_{tn} + \mathrm{SU}_{tn} + \left(a_n + b_n P_{tn} + c_n P_{tn}^2\right)\right) \qquad (5.2)$$

where r_n and R_{tn} are the reserve cost and the online spinning reserve, respectively. SU_{tn} is the conventional generator startup cost. The quadratic function represents the generation cost of the conventional generator. The flowchart in Figure 5.1 explains the process of the ESS sizing algorithm. The proposed algorithm begins by determining the minimum required BESS capacity in both discharging ($E_{\mathrm{dis}}^{\min}$) and charging ($E_{\mathrm{ch}}^{\min}$) modes. In discharging mode, the BESS should be able to compensate the maximum power shortage in the microgrid system. The maximum power shortage is defined as the difference between the electrical demand and the total maximum generation of all the units within the microgrid at time t. Meanwhile, the BESS should be able to absorb the maximum excess power in the microgrid when the minimum total generation exceeds the demand. The minimum required capacity of the BESS ($E_{\mathrm{BESS}}^{\min}$) is chosen as the higher value of the two calculated minimum capacities taking into account the discharging efficiency (η_d) and the charging efficiency (η_c). Fuzzy logic techniques can be used to find the minimum BESS capacity. Initially, the BESS rated energy is set to be equal to $E_{\mathrm{BESS}}^{\min}$ and the unit commitment problem is solved based on this value. The process is then repeated with increasing the BESS rated energy by an incremental value (ΔE) until a specified maximum value ($E_{\mathrm{BESS}}^{\max}$) is reached. The optimal size of the BESS would be the size that yields the minimum unit commitment cost. Mixed nonlinear integer programming is used to solve the problem. However, when the ESS investment cost as well as the ability of the microgrid to operate in parallel to main grid are considered in the problem, then the objective function will be to minimize

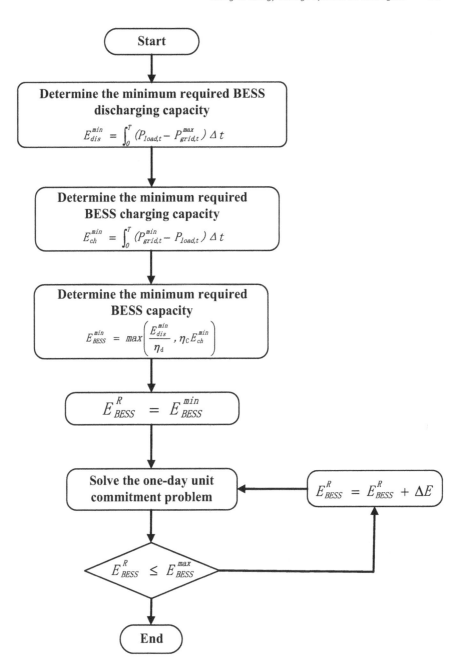

Figure 5.1 Proposed BESS sizing algorithm.

the total cost and maximize the market profit [7]. These two objectives are stated as:

$$TC = TCPD + TUCC \qquad (5.3)$$

$$TB = MB - TCPD \qquad (5.4)$$

where,

TC = Total cost for islanded microgrid operation mode
TCPD = Total cost per day of Battery ESS installed
TB = Total benefit for grid-connected microgrid operation mode
MB = Market benefit
TUCC = Total unit schedule cost

A trade-off is obtained between total cost and total benefit to obtain optimal ESS size.

One of the important applications of ESS, especially in islanded microgrid, is voltage and frequency support. When the microgrid is operating in islanded mode, the total load demand within the microgrid must be satisfied by local generation. However, with high renewable sources penetration, which is intermittent in nature, it is challenging to maintain the balance between demand and generation at all times. Moreover, the sudden transition from grid-connected to islanded mode causes a power unbalance in the microgrid system. Failure to maintain the power balance in the microgrid leads to frequency deviation that may cause, in the worst case scenario, the microgrid system to collapse. Thus, integrating an ESS is indispensable for an islanded microgrid. BESSs in general are characterized as fast response energy storage systems. This makes them suitable for voltage and frequency regulation applications. Determining the optimal size of the BESS for primary frequency control is one of the important subjects for ESS sizing in microgrids [4,9].

In [4], a PSO technique is implemented to find the optimal size of a BESS in combination with load shedding to regulate the microgrid frequency when it is operating in islanded mode. When the microgrid is disconnected from the main grid due to any disturbance, the microgrid load must be supplied by local generation, which may not be sufficient to meet the load at that moment. When local load is more than generation, the system frequency will drop thus affecting system stability. Normally, load shedding is used to restore the frequency to its nominal value. However, application of BESS together with load shedding can greatly improve frequency regulation capacity better serve the load and reduce the operational cost. The objective function in this method is to minimize the power of the BESS subject to frequency regulation constraints. In [8], a new approach that takes advantage of the overloading characteristics of a BESS is presented. The BESS overloading characteristics include overloading capacity and permissible overloading duration. Since the primary frequency control requires only a small but rapid amount of power, it is possible to utilize the overloading characteristics of the BESS to provide this required burst of power. In this way, it is possible to choose a small-sized BESS to regulate the frequency in the microgrid thus reducing its related cost.

In [10], a mathematical method for determining the size of ESS installed in order to meet a critical load reliability requirement is investigated. If PC denotes the critical load's power that must be supplied during the main power source outage events (S_F), the ESS power rating must be equal to this value. However, the energy value is based on the required time duration that is needed for the ESS to supply the load when the main power source is out for any reason (t_A). This value can be found from the following equations:

$$
\begin{aligned}
P\{L\} &= P\{\{R > t_A\} \cap S_F\} \\
&= P\{R > t_A | S_F\} P\{S_F\} \\
&= \left(\int_{t_A}^{\infty} f_R(r)\mathrm{d}r \right) P\{S_F\}
\end{aligned}
\tag{5.5}
$$

where $P\{L\}$ represents the probability that the critical load is not being supplied. R is a random variable denoting the outage time of the main power source. $f_R(r)$ represents the probability density function of R. Now, it is assumed that A_0 represents the availability of the main power source in the system. The objective of installing the ESS is to increase this value to A_1. Thus, a new factor called unavailability reduction ratio can be defined as:

$$
\alpha = \frac{1 - A_1}{1 - A_0}
\tag{5.6}
$$

Note that $P\{L\}$ and $P\{S_F\}$ can be defined in terms of the old and new availability factors (A_0) and (A_1) respectively as follows:

$$
P\{L\} = 1 - A_1
\tag{5.7}
$$

$$
P\{S_F\} = 1 - A_0
\tag{5.8}
$$

Thus, (5.5) can be rewritten as

$$\int_{t_A}^{\infty} f_R(r)\mathrm{d}r = \alpha \qquad (5.9)$$

This equation is considered as the basic relationship based on which ESS size can be found. The outage time of the main power source is exponentially distributed when its failure rate is considered constant. In this case, $f_R(r)$ can be represented as:

$$f_R(r) = \frac{1}{\bar{r}} \exp\left(\frac{-r}{\bar{r}}\right), r \ge 0 \qquad (5.10)$$

where $\bar{r}$ represents the mean of R. The value of t_A can be calculated from:

$$t_A = -\bar{r} \ln \alpha \qquad (5.11)$$

Thus, the installed ESS energy rating is equal to $t_A \times PC$. If the ESS availability (A_S) is to be considered, the time for which the ESS needs to supply the load is increased to t_S, which can be defined as:

$$t_S = \frac{t_A}{A_S} \qquad (5.12)$$

Moreover, non-constant primary generator failure rates, where R is represented by other distributions such as Weibull or lognormal, are also discussed in [9].

5.3 CASE STUDY: ENERGY STORAGE SIZING IN MICROGRID

5.3.1 Problem Formulation

One of the most common methods to formulate the optimization problem for sizing ESS is mixed integer linear programming (MILP). In this section, MILP is implemented to formulate the problem of determining both the optimal power rating and energy rating of an ESS integrated to a microgrid in order to minimize the microgrid total cost, which includes its operational cost as well as the investment cost of the ESS. The operational cost consists of the generation cost of the dispatchable units within the microgrid as well as the cost of energy interchange between the microgrid and the main grid. The energy storage investment cost depends on its power and energy ratings. This cost comprises power initial cost in $/kW, energy initial cost in $/kWh, operating and maintenance cost in $/kW, conversion system cost in $/kW, and disposal cost in $/kW. Based on those costs, the objective function of optimization problem can be defined as:

$$PC_B P_{ESS}^R + EC_B C_{ESS}^R + \sum_i \sum_t (F(P_{it})I_{it+}SU_{it} + SD_{it}) + \sum_t \rho_t P_{M,t}$$

(5.13)

Here PC_B and EC_B are the power and energy related cost, respectively. P_{ESS}^R and C_{ESS}^R are power and energy ratings of ESS, respectively. The third term represents the generation cost of the dispatchable units within the microgrid, which includes the fuel cost and the cost of starting up (SU) or shutting down (SD) the generation units. The last term accounts for the cost/benefit of buying/selling energy from/to the main grid. When the power is imported from the main grid, $P_{M,t}$ will be positive. Note that sign of $P_{M,t}$ will be negative when the power is exported to the main grid. ρ_t is the electricity price at the point of connection between the microgrid and main grid. t represents the time in hour. Equation (5.13) is solved subject to system constraints, generation unit constraints and energy storage constraints. Those constraints are discussed in detail in the following sections.

5.3.2 System Constraints
The system constraints ensure the power balance within the microgrid as well as limit the power exchanged between the microgrid and the main grid. They can be defined as follows.

$$\sum_{i\in\{G,W\}} P_{it} + \sum_t P_{ESS,t} + P_{M,t} = D_t \quad \forall t \qquad (5.14)$$

$$-P_M^{max} \le P_{M,t} \le P_M^{max} \quad \forall t \qquad (5.15)$$

Power balance Eq. (5.14) denotes that the summation of power generated from local distributed resources, the power to or from the storage system, and the power from or to the main grid satisfies the load in each hour. G and W denote the number of dispatchable units and the number of renewable energy generators, respectively. The exchanged power with the main grid is limited by the capacity of the line connecting them as given in Eq. (5.15).

5.3.3 Generation Units Constraints

The dispatchable units have some physical constraints that must be considered when the optimization problem is formulated. The generation unit's output power is limited by maximum and minimum values (5.16). Moreover, the variation in the output power between two successive hours is limited by ramp up and ramp down limits (5.17) and (5.18). When the generation unit starts up, it stays on for minimum time (5.19). Similarly, once the unit shuts down, it must stay off for a specific minimum time (5.20). Those constraints can be expressed as follows:

$$P_i^{\min} I_{it} \le P_{it} \le P_i^{\max} I_{it} \quad \forall i \in G, \forall t \tag{5.16}$$

$$P_{it} - P_{i(t-1)} \le \text{UR}_i \quad \forall i \in G, \forall t \tag{5.17}$$

$$P_{i(t-1)} - P_{it} \le \text{DR}_i \quad \forall i \in G, \forall t \tag{5.18}$$

$$T_{it}^{\text{ON}} \ge \text{UT}_i (I_{it} - I_{i(t-1)}) \quad \forall i \in G, \forall t \tag{5.19}$$

$$T_{it}^{\text{OFF}} \ge \text{DT}_i (I_{i(t-1)} - I_{it}) \quad \forall i \in G, \forall t \tag{5.20}$$

The binary variable I_{idh} represents the unit commitment state. When the unit is on, the value of I_{it} is 1, otherwise, it is 0. UR_i and DR_i are the ramp up and ramp down limits, respectively. UT_i and DT_i are the minimum up and down time for the generation unit. Satisfying those constraints ensures safe operation of the generation unit.

5.3.4 Energy Storage System Constraints

The following equations model the energy storage system operation:

$$0 \le P_{\text{ESS},t}^{\text{dis}} \le P_{\text{ESS}}^{R} u_{1,t} \quad \forall t \tag{5.21}$$

$$-P_{\text{ESS}}^{R} u_{2,t} \le P_{\text{ESS},t}^{\text{ch}} \le 0 \quad \forall t \tag{5.22}$$

$$P_{\text{ESS},t} = P_{\text{ESS},t}^{\text{dis}} - P_{\text{ESS},t}^{\text{ch}} \quad \forall t \tag{5.23}$$

$$u_{1,t} + u_{2,t} \le 1 \quad \forall t \tag{5.24}$$

$$C_{\text{ESS},t} = C_{\text{ESS},(t-1)} - \frac{P_{\text{ESS},t}^{\text{dis}}}{\eta_{\text{ESS}}} - P_{\text{ESS},t}^{\text{ch}} \quad \forall t \tag{5.25}$$

$$0 \le C_{\text{ESS},t} \le C_{\text{ESS}}^{R} \quad \forall t \tag{5.26}$$

Table 5.1 Generation Units Technical Characteristics				
Unit Number	Min.–Max. Capacity (MW)	Min. UP/ Down time (h)	Ramp Up/Down Rate (MW/h)	Cost Coefficient ($/MWh)
1	0.8–5	3	2.5	27.7
2	0.5–5	3	2.5	61.3

The ESS charging and discharging power are limited by the rated power (5.21) and (5.22). The ESS power, $P_{ESS,t}$, is negative in charging mode, positive in discharging mode, and zero in idle mode. The binary variables $u_{1,t}$ and $u_{2,t}$ indicate the discharging and charging states, respectively. Equation (5.24) implies that the ESS cannot be simultaneously charged and discharged. The energy stored in the ESS at each hour is determined by Eq. (5.25) and limited by Eq. (5.26). For more realistic consideration, the ESS state of charge constraint $(SOC^{min} \leq SOC \leq SOC^{max})$ should be included.

5.3.5 Numerical Simulation

A microgrid is used to perform a case study for the aforementioned method. It consists of four gas generators, a photovoltaic (PV) array, a wind turbine, an energy storage system, and local loads as shown in Figure 5.2. The technical characteristics of the dispatchable gas generators are shown in Table 5.1. The PV array power rating is 1.5 MW and the wind generator power rating is 1 MW. The hourly output power of the renewable sources, the microgrid local load, and the market price of electricity at the point of common coupling (PCC) are obtained from historical data. One-week generation profiles of the renewable sources, microgrid load, and the electricity price are shown in Figures 5.3–5.5. The objective is to determine the optimal size of the energy storage system that minimizes the microgrid total cost. The initial power and energy costs of the considered ESS are 40 $/kW/year and 11 $/kWh/year [2], respectively. Note that the operating and maintenance cost, the conversion system capital cost, as well as the ESS disposal cost are included in the initial power rating cost. In addition, the given costs are annualized over the ESS life time. It is assumed that the ESS round trip efficiency is 85%. The exchanged power with the main grid is limited by the capacity of the line connecting the microgrid to the main grid. This line capacity is assumed to be 6 MW in this simulation.

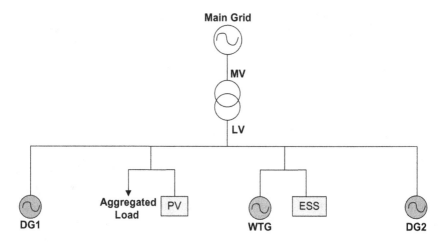

Figure 5.2 Microgrid configuration.

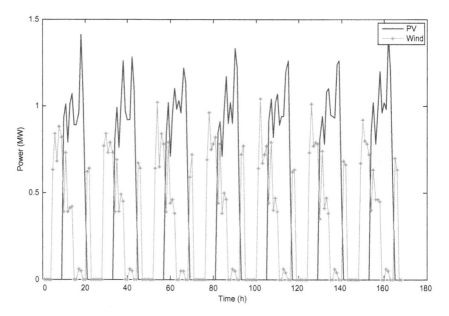

Figure 5.3 Renewable energy resources output power for one week.

5.3.6 Results and Discussions

To illustrate the economic benefit of integrating an ESS into a microgrid, the total cost of the microgrid is found first without any ESS. Since there is no ESS in the system, the total cost includes only the cost of generation units and power exchanged with the main grid to supply the local load. This total cost is found to be equal to

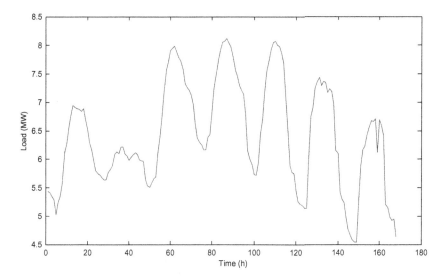

Figure 5.4 Microgrid local load for one week.

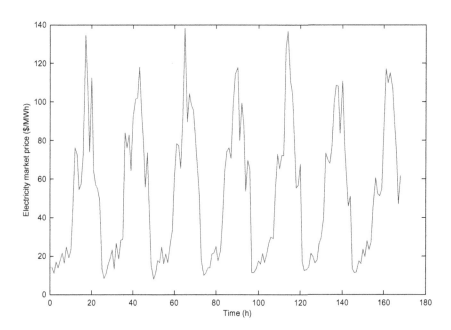

Figure 5.5 Market price of electricity for one week.

224,528 $/year. The generation unit cost is 1,817,433 $/year and the cost exchanged power with the main grid is −1,592,905 $/year. The negative sign means that the benefit made by selling energy to the main grid is higher than the cost of buying energy from the main grid.

When the ESS is added to the microgrid, it is found that the optimal power and energy ratings are 750 kW and 6 MWh, respectively. The microgrid total cost is reduced to 219,078 $/year. This cost comprises an ESS investment cost (96,000 $/year), generation cost (1,739,008 $/year), and exchanged power with main grid cost (−1,615,930 $/year). Table 5.2 summarizes the microgrid costs with and without an ESS. Figure 5.6 explains how the microgrid costs change with respect to the ESS energy rating while the power rating is

Table 5.2 Microgrid Costs				
Cases	Dispatchable Units Generation Cost ($/year)	Exchanged Power with Main Grid Cost ($/year)	ESS Investment Cost ($/year)	Total Cost ($/year)
Without ESS	1,817,433	− 1,592,905	0	224,528
With optimal sized ESS	1,739,008	− 1,615,930	96,000	219,078

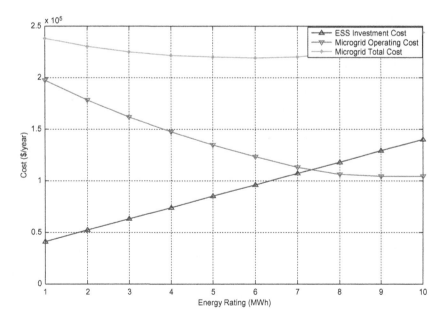

Figure 5.6 Optimal ESS sizing in microgrid application.

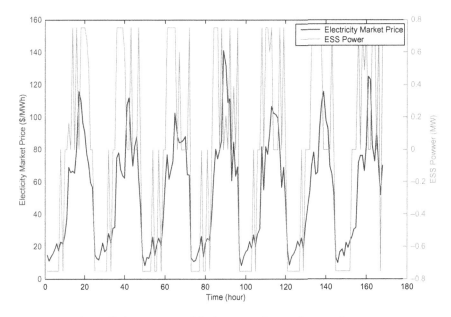

Figure 5.7 ESS charging/discharging power and the electricity market price for one week.

kept constant at the optimal value. It is clear that the minimum total cost can be found approximately at 6 MWh. The ESS charging and discharging power is shown in Figure 5.7. As can be seen, the ESS is charged when the market price is low and discharged when the price is high. In this way, the ESS makes a profit from selling the energy to the main grid during the high price period. This operation is known as energy arbitrage. The hourly exchanged power with the main grid is depicted in Figure 5.8. It is clear that as the electricity price reaches its maximum value, the amount of power sold to the main grid reaches its maximum value as well.

In order to examine the impact of an ESS size on the total cost of the microgrid, the optimization problem is resolved with a variety of ESS sizes. The results are depicted in Figures 5.9–5.11. As can be seen from Figure 5.9, the ESS investment cost increases linearly as the power and energy ratings of the ESS increase. By increasing the storage size, the operating cost of the system reduces, as seen in Figure 5.10. A higher size of the storage system can store more energy at off-peak hours and thus produce more energy at peak hours, which provides higher economical benefits for the microgrid system. However, for a 1 MW power rating the operating cost reaches a saturated point after which it is almost constant. This is because with such

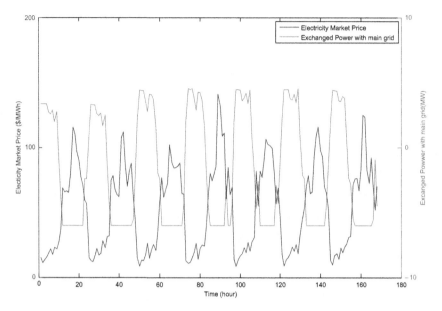

Figure 5.8 Power exchanged with the main grid for one week.

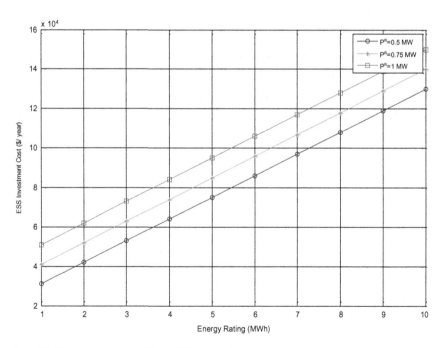

Figure 5.9 ESS investment cost at different ESS power and energy ratings.

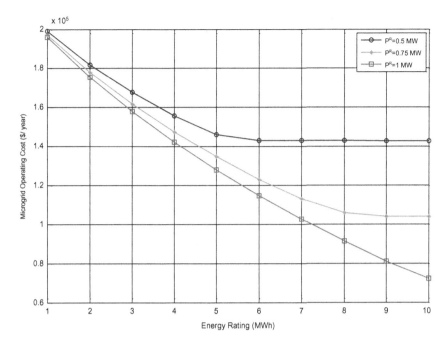

Figure 5.10 Microgrid operating cost at different ESS power and energy ratings.

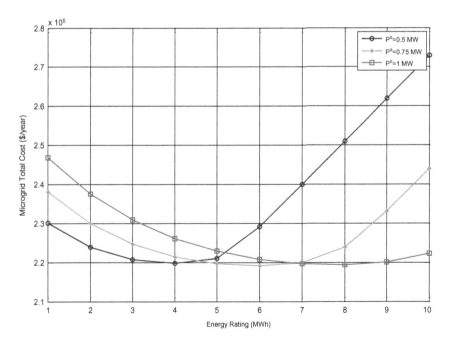

Figure 5.11 Microgrid total cost at different ESS power and energy ratings.

small power rating, the benefit of selling power to the main grid is limited. The summation of the ESS investment cost and the operating cost results in the microgrid total cost as shown in Figure 5.11. It is clear that for each power rating, there is a different optimal energy rating value for the ESS. Thus, it is important to consider both the power and energy ratings of the ESS when sizing problem is solved.

REFERENCES

[1] Bahramirad S, Daneshi H. Optimal sizing of smart grid storage management system in a microgrid. In: Innovative smart grid technologies (ISGT), 2012 IEEE PES. January 2012. p. 1–7.

[2] Bahramirad S, Reder W, Khodaei A. Reliability-constrained optimal sizing of energy storage system in a microgrid. IEEE Trans Smart Grid 2012;3(4):2056–62.

[3] Hartono B, Budiyanto Y, Setiabudy R. Review of microgrid technology. In: 2013 international conference on QiR (Quality in Research). June 2013. p. 127–32.

[4] Kerdphol T, Qudaih Y, Mitani, Y. Battery energy storage system size optimization in microgrid using particle swarm optimization. Innovative smart grid technologies conference europe (ISGT-Europe), 2014 IEEE PES, 12–15 October 2014. p. 1–6.

[5] Nguyen T, Crow M, Elmore A. Optimal sizing of a vanadium redox battery system for microgrid systems. Sustainable Energy, IEEE Transactions on, vol. PP, no. 99, p. 1–9.

[6] Ross M, Hidalgo R, Abbey C, Joos G. Analysis of energy storage sizing and technologies. In: Electric power and energy conference (EPEC), 2010 IEEE. August 2010. p. 1–6.

[7] Chen SX, Gooi HB. Sizing of energy storage system for microgrids. 2010 IEEE 11th international conference on probabilistic methods applied to power systems (PMAPS), 14–17 June 2010. p. 6–11.

[8] Chen SX, Gooi HB, Wang AQ. Sizing of energy storage for microgrids. IEEE Trans Smart Grid 2012;3(1):142–51.

[9] Aghamohammadi MR, Abdolahinia H. A new approach for optimal sizing of battery energy storage system for primary frequency control of islanded Microgrid. Int J Electr Power Energy Syst January 2014;54:325–33.

[10] Mitra J, Vallem MR. Determination of storage required to meet reliability guarantees on island capable microgrids with intermittent sources. IEEE Trans Power Syst November 2012;27(4).

[11] Khodaei A. Microgrid optimal scheduling with multi-period islanding constraints. IEEE Trans Power Syst 2014;29(3):1383–92.

Printed in the United States
By Bookmasters